SketchUp 2013 for Interior Designers

Daniel John Stine

SDC
Publications

SDC Publications

P.O. Box 1334
Mission, KS 66222
913-262-2664
www.SDCpublications.com
Publisher: Stephen Schroff

ISBN-13: 978-1-58503-838-1
ISBN-10: 1-58503-838-5

Printed and bound in the United States of America.

Foreword:

This book has been written with the assumption that the reader has no prior experience using Trimble® **SketchUp®** (formerly *Google SketchUp*). With this book, the reader will be able to describe and apply many of the fundamental principles needed to develop compelling SketchUp models.

The tutorials introduce the reader to SketchUp, an easy to use 3D modeling program geared specifically towards interior design. Several pieces of furniture are modeled. The process is broken down into the fundamental concepts of 2D line work, 3D extraction, applying materials and printing.

Rather than covering any one feature or workflow in excruciating detail, this book aims to highlight many topics typically encountered in practice. Many of the tutorials build upon each other so the reader has a better understand of how everything works and ends with a greater sense of accomplishment.

In addition to "pure" SketchUp tutorials, which comprises most of the text, the reader will also enjoy these "extended" topics:

- Introduction to **LayOut**; *an application which comes with SketchUp Pro*
- Manufacturer specific paint colors and wallcoverings
- Manufacturer specific furniture
- Manufacture specific flooring
- Photorealistic rendering using **V-Ray for SketchUp**
- Working with **AutoCAD** DWG files
- Working with **Revit**; *including how to bring SketchUp models into Revit*

Although the book is primarily written with a classroom setting in mind, most individuals will be able to work through the tutorials on their own and benefit from the tips and tricks presented.

For a little inspiration, this book has several real-world SketchUp project images throughout. Also, a **real-world project is provided** to explore and it is employed in the book to develop a **walkthrough animation**. ENJOY!

About the Author:

Dan Stine is a registered Architect with twenty-two years of experience in the architectural field. He currently works at LHB (a 240 person multidiscipline firm; www.LHBcorp.com) in Duluth, Minnesota as the CAD/BIM Administrator, providing training, customization and support for two regional offices. Mr. Stine has presented at Autodesk University (au.autodesk.com) and at the Revit Technology Conference (revitconference.com).

Dan has worked in a total of four firms. While at these firms, he has participated in collaborative projects with several other firms on various projects (including Cesar Pelli, Weber Music Hall – University of Minnesota - Duluth). Dan is a member of the *Construction Specification Institute* (CSI) and the *Autodesk Developer Network* (ADN) and also teaches *Autodesk Revit* to interior design students at North Dakota State University (NDSU). He previously taught *AutoCAD* and *Autodesk Revit* classes for 12 years at Lake Superior College, in the Architectural Technology program. Additionally, he is a certified Construction Document Technician (CDT). Mr. Stine has also written the following textbooks (published by SDC Publications):

- *Interior Design using Hand Sketching, SketchUp and Photoshop (with co-author Steven H. McNeill)*
- *Interior Design using Revit Architecture 2014*
- *Residential Design using Revit Architecture 2014*
- *Commercial Design using Revit Architecture 2014*
- *Design Integration using Revit 2014 (Architecture, Structure and MEP)*
- *Residential Design using AutoCAD 2014*
- *Commercial Design using AutoCAD 2013*
- *Chapters in Architectural Drawing (with co-author Steven H. McNeill, AIA, LEED AP)*

Mr. Stine likes to spend time with his family, camping, watching movies, fishing and much more! He also enjoys riding his bicycle along the shoreline of Lake Superior, and to work when there is no snow on the ground. The total ride distance is only 4 miles each way, but the elevation gain is over 500 feet. Luckily the uphill part is after work!

You can contact the publisher with comments or suggestions at service@sdcpublications.com

Thanks:

I could not have done this without the support from my family; Cheri, Kayla & Carter.

Many thanks go out to Stephen Schroff and SDC Publications for making this book possible!

Table of Contents

Chapter 1
Introduction to SketchUp

What is SketchUp® used for?

It might be easier to answer what SketchUp is not used for. SketchUp is an all-purpose 3D modeling tool. The program is primarily developed around architectural design but it can be used to model just about anything. The program's relative ease of use and low cost (the basic package, SketchUp Make, being free) makes it a very popular tool within the AEC design community.

It should be pointed out that SketchUp was recently sold by Google to Trimble. The software is still being developed by the same folks in Colorado; they just get their paychecks from someone else. Trimble is a leading manufacturer of survey equipment and GIS software.

Why use SketchUp?

As just mentioned, it is easy to use and comes in a free version! It is a simple way to quickly communicate your design ideas to clients or prospective employers. Not only can you create great still images, SketchUp also is able to produce walk-thru 245videos!

When creating interior designs using SketchUp you have access to a massive amount of content with via **Trimble's 3D Warehouse**. You can take a peek now if you want: **http://sketchup.google.com/3dwarehouse/**. Keep in mind that this link may change at some point, due to the recent acquisition by Trimble.

SketchUp versus other Applications?

There are several other 3D modeling applications (aka programs) on the market which compete with SketchUp to varying degrees. Every program has its strengths and weaknesses when compared to another. SketchUp is mainly geared towards concept designs rather than construction documentation. Its ability to quickly develop and present the designer's ideas makes it very popular. However, it is not very good at adding notes and dimensions, but it is getting better at that with each new version. So at some point the SketchUp design typically needs to be exported to a CAD format, such as Autodesk's DWG format and finished in AutoCAD or a similar program.

Another popular modeling approach is Building Information Modeling (BIM). SketchUp is not a BIM application. However, SketchUp can still have a place in the BIM workflow. An application such as Autodesk's Revit® Architecture does have several SketchUp-like features, but many designers prefer the simplicity and limited scope (i.e., SketchUp is designed to do one thing, and it does that one thing very well – similar to Five Guys burgers!). Plus, Revit can import SketchUp models.

SketchUp is primarily a face-based modeling program, as opposed to a solid modeling program. This has its pluses and minuses. It is great for concept modeling as it keeps the size and complexity of the model down to a minimum. This allows the designer to quickly spin around the model and zoom in and out, whereas a solids-based model could take nearly a minute to spin around where SketchUp could do it in seconds. This is all relative to the project size – a small simple project would not be a problem in either case, but a 200,000 square foot school or hospital likely would be. Speed also depends on the power of your computer of course; processor speed, RAM (quantity and speed), video card, video card memory, video card software driver. Notice the emphasis on video card? This is because SketchUp is a graphics intense program, and benefits from high quality video cards when spinning around the model, while displaying textures (e.g. an image of brick) and generating shadows.

This may be considered debatable by some SketchUp experts, but a solids-based model is generally easier to make changes to when the model is complex. SketchUp Pro now has the ability to create some solids so the playing field is leveling out.

One of the drawbacks to face-based modeling is the designer cannot schedule information from the model such as cubic foot of material for a concrete wall. Luckily nobody really cares about that in the early stages of design! A similar problem is that elements in section look hollow in SketchUp whereas in a solids-based program they would not. This can be seen in the example below:

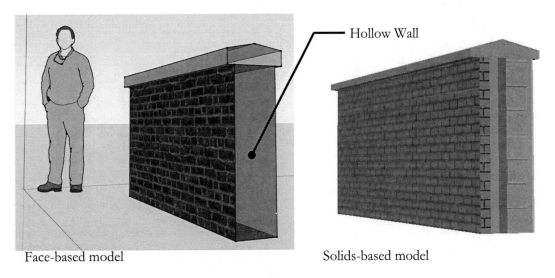

Face-based model   Solids-based model

FIGURE 1.1 SketchUp is a face-based program, not solids-based.

The left image shows a masonry wall cut using the section tool. Notice the wall appears hollow. SketchUp does not know this is a masonry wall, or even a wall for that matter. The image on the right is a BIM model – notice the solid core.

As already mentioned, it is possible to export a SketchUp model to a CAD program such as AutoCAD. This format does not work too well in a BIM program such as Revit – but it can be used to varying degrees. It is also possible to export a BIM model to a format that can be imported into SketchUp – thus allowing some of SketchUp's tools to be used; hand sketch effect, easy navigation and simple animation setup and creation.

SketchUp Pro versus SketchUp Make (the Free version)?

The free version of SketchUp, which is called **SketchUp Make**, is very powerful and can model just about anything. The Pro version, called **SketchUp Pro**, costs $495 at the time of this printing, has several advanced features such as:

- Technical support
- Solid modeling
- Import and Export AutoCAD DWG files
- Layout 2013
    - A separate program used to compose multiple views of the same model on a page
    - Add notes and dimensions
    - More printing options
- Style Builder
    - A separate program used to transform your model into a unique hand drawing.

You can see a more detailed comparison of the free versus pro version at the following web address:

- http://www.sketchup.com/intl/en/product/whygopro.html

For most design firms the pro version is a must, if just for the ability to use and export AutoCAD DWG files. For example, you might import a 2D DWG file provided by the client and use that line work to quickly start modeling the existing conditions. SketchUp also has a network license option which a company can purchase. This allows a firm to have the software installed on everyone's computers, but the number of people who can access the program, at any given time, is limited to the number of network licenses the company owns. So, if your firm has six licenses available on the server, the seventh person gets a denial message. That person can keep trying until a license becomes available or, better, make a few calls or send out an email to see if someone can get out!

Given the introductory nature of this textbook, we will mostly cover the tools and techniques found in the free version of SketchUp. A few Pro features will be covered; these steps will be clearly marked. If you only have *SketchUp Make*, simply skip the "Pro" sections.

Mac versus PC?

SketchUp has been designed to work on either the Apple Macintosh or Microsoft Windows based computer system. Most Architectural and Interior Design offices tend to favor the PC due to cost and general availability of other programs geared towards the industry. However, there is not much that cannot be done on a Mac, especially with its ability to run Windows when using Mac's *Boot Camp* or a virtual environment (e.g. *Parallels*).

All screen shots in this book are from a PC running Windows 7 64bit. If the reader is using a Mac or another version of Windows there might be slight differences in some screen shots. However, the main SketchUp *User Interface* should be the same; the User Interface is covered next.

Warning!

Because you are using this book, you are obviously interested in Interior Design and SketchUp is great for that. But, just so you know (and to provoke a little smile), there are some things SketchUp is not allowed to be used for. The following is an excerpt from the End User License Agreement you are required to agree to before downloading and using the software:

NONE OF THE SOFTWARE IS INTENDED FOR USE IN THE OPERATION OF NUCLEAR FACILITIES, LIFE SUPPORT SYSTEMS, EMERGENCY COMMUNICATIONS, AIRCRAFT NAVIGATION OR COMMUNICATION SYSTEMS, AIR TRAFFIC CONTROL SYSTEMS, OR ANY OTHER SUCH ACTIVITIES IN WHICH CASE THE FAILURE OF THE SOFTWARE COULD LEAD TO DEATH, PERSONAL INJURY, OR SEVERE PHYSICAL OR ENVIRONMENTAL DAMAGE.

SketchUp Viewer

A streamlined viewer is available for download from the SketchUp website. This is great to recommend to clients who might want to view the model in 3D on their screen rather than looking at a still image file (e.g. JPG or PNG). The viewer does not have any model editing tools, so there are less options for the person reviewing the model to learn. In addition to viewing the model, one can also print. If you have SketchUp Make or SketchUp Pro you do not need SketchUp Viewer. The free viewer can be download from: http://www.sketchup.com/products/sketchup-viewer.

SketchUp Extensions

Although SketchUp can do a lot, sometimes you just want it to do more. For that, one can turn to the Extension Warehouse to see if someone created an add-in for a specific need. Here you will find tools to perform energy analysis, photorealistic renderings and much more. Some of these are free while others are not. You can learn more about this via this link: http://extensions.sketchup.com/

Staying in the loop

Be sure to follow the official SketchUp blog to learn about the latest updates, tips and tricks. You can also read about how others are using the product. http://sketchupdate.blogspot.com/

Project Example: Waiting Area.   *Image courtesy of LHB; www.LHBcorp.com*

Chapter 1 Review Questions

The following questions may be assigned by your instructor as a way to assess your knowledge of this chapter. Your instructor has the answers to the review questions.

1. SketchUp is a face-based 3D modeling program. (T/F)

2. SketchUp is strictly a 2D drafting program. (T/F)

3. When using SketchUp you access content via 3D Warehouse. (T/F)

4. Custom programs can be created to extend the capabilities of SketchUp. (T/F)

5. SketchUp was not primarily created for architecture. (T/F)

6. The free version of SketchUp is called SketchUp _____. (T/F)

7. Name one of the authors of the official SketchUp blog:

   _____   _____.

8. SketchUp only works with Windows, not Apple computers. (T/F)

Chapter 2
Overview of the SketchUp User Interface

The first step in learning any new computer program is figuring out the **User Interface** (UI). SketchUp has a traditional user interface, consisting of menus across the top, toolbars that can float or be docked to a side, a status bar across the bottom and a large area in the middle to do your design work. SketchUp has chosen not to implement the *Ribbon* as a number of other software makers have (e.g., Microsoft and Autodesk).

The image on the next page highlights the primary components of the *User Interface*. A few of the items identified are not really considered a part of the UI, but help paint a better overall picture for the new user. Take a little time to familiarize yourself with the terms presented, as they will be used throughout this textbook.

Project Example: Custom Casework.   *Image courtesy of LHB; www.LHBcorp.com*

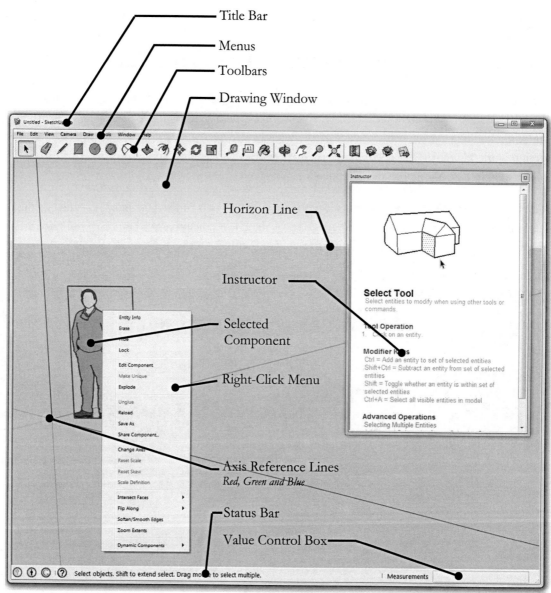

Title Bar

Menus

Toolbars

Drawing Window

Horizon Line

Instructor

Selected Component

Right-Click Menu

Axis Reference Lines
Red, Green and Blue

Status Bar

Value Control Box

FIGURE 2.1 SketchUp User Interface

Title Bar

The program title bar displays the name of the file currently being worked on, *TFDM Office Building Remodel.skp* in this case. And just in case you forgot, the name of the program you are using is listed to the right of the file name: *SketchUp*, right? On the far right are the typical controls for the application's window: minimize, maximize/restore down, and close.

FIGURE 2.2 Title Bar

Menus

Below the title bar are several pull-down menus; these are the words across the top (File, Edit, View, etc.). When clicked on, these menus reveal a list of commands. The menus are a way to break the list of commands down into smaller, task specific lists. A *menu* is closed when a command is selected or the **Esc** key is pressed.

Notice in the image provided (Figure 2.3), the **Camera** menu is expanded. Some items have a check mark on the left to show that item is active (projection type in this case). Also, some items in the list are fly-out menus. A fly-out menu can be identified by the black arrow pointing to the right (**Standard Views**, for example). Hovering over a fly-out menu item reveals another sub-set of menu options. Whenever a command or toggle has a keyboard shortcut, it is listed on the right. Finally, any command or toggles which are not relevant to the current tool or drawing will be grayed out to avoid any confusion.

FIGURE 2.3 Menus

When reference is made to a command within the menu system it will be shown as such:

Camera → Parallel Projection

This means: click the **Camera** menu and then click the **Parallel Projection** command in the list as shown in the image to the right (Figure 2.4).

FIGURE 2.4 Menu selection

Toolbars

Toolbars are a favorite for most SketchUp users as they provide small graphical images and are only a single click away. When SketchUp is first started (after being installed), only the *Getting Started* toolbar is showing (Figure 2.5). Additional toolbars can be toggled on and off via **View → Toolbars**. Whenever a toolbar is visible on the screen, it can be dragged so it is "docked" along the perimeter of the drawing window, or it can "float" anywhere on the screen.

The book will give specific instructions when certain toolbars are required. It is recommended that toolbars only be turned on when instructed to minimize any possible confusion and so the reader's screen matches the images in the book.

FIGURE 2.5 Toolbars

Drawing Window

The drawing window is, of course, where all the modeling is done! Using various tools from the menus, toolbars and keyboard shortcuts, you create and interact with your model in the drawing window.

Status Bar (and Value Control Box)

The *Status Bar* is found across the bottom of the application (Figure 2.6). On the far left are three small round icons; hover your cursor over them to see what they are; they are not critical at this time. Next you have a circle with a question mark which toggles the *Instructor* visibility on and off (see the next topic for more on this feature). The next section, to the right of the instructor toggle, provides prompts for any command you are currently using. The example shown is letting you know SketchUp expects you to pick a point in the model to define one of the corners while using the *Rectangle* tool. Finally, on the far right hand side of the *Status Bar* is the **Value Control Box**. This box shows the length or size of an object being drawn. It is not necessary to spend a lot of time moving the mouse into just the right location so the dimension reads correctly, as you can more quickly (and accurately) type this information in (either before or after picking your last point).

FIGURE 2.6 Status Bar

Instructor

The *Instructor* is not necessarily part of the *User Interface*, but it automatically appears on the screen when SketchUp is opened. This feature is intended to help new users understand how to use various tools. For example, when you select the **Rectangle** tool the *Instructor* provides an animated graphic and steps on how to sketch a rectangle (Figure 2.7). This feature compliments this book in that it will remind you how various tools are used. This book works through many of SketchUp's commands in a systematic way, and once a command is covered it is not typically covered again in as much detail.

Once you become familiar with SketchUp, you will want to turn off the *Instructor* in order to free up more screen space. This can be done via **Window → Instructor**; clicking this toggles the *Instructor* on and off. Also, as mentioned in the previous section, the question mark icon on the *Status Bar* will turn the *Instructor* on, if off – and it will also minimize it if already on.

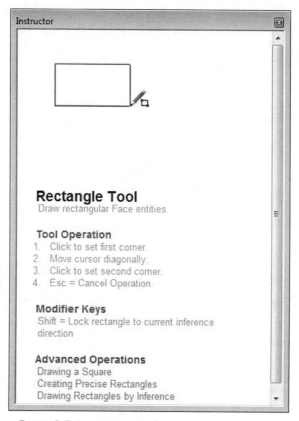

FIGURE 2.7 Instructor window

If you have a dual monitor computer system (see example on next page), the *Instructor* can be moved to the second monitor to increase the usable portion of the *Drawing Window*.

Dual screen setup

Right-Click Menu

SketchUp allows you to right-click on something to both select it and present a contextual pop-up menu which provides quick access to tools used to manipulate the selected component, line/edge or face. Notice in Figure 2.8a that there are fewer options to choose from when an edge is selected than for a component (Figure 2.8b). Also, some options, in the right-click menu, have a black triangle pointing to the right. Hovering over these reveals a sub-menu with additional tools, similar to the pull-down menus.

Clicking a tool from the right-click menu or pressing the **Esc** key closes the right-click menu.

Entity Info
Erase
Hide

Select ▶
Soften
Divide
Zoom Extents

FIGURE 2.8A
Right-click menu:
Edge Selected

FIGURE 2.8B ⟶
Right-click menu:
Component Selected

Entity Info
Erase
Hide
Lock

Edit Component
Make Unique
Explode

Unglue
Reload
Save As
Share Component...

Change Axes
Reset Scale
Reset Skew
Scale Definition

Intersect Faces ▶
Flip Along ▶
Soften/Smooth Edges
Zoom Extents

Dynamic Components ▶

Project Example: Retail Counter.

Image courtesy of LHB; www.LHBcorp.com

Chapter 2 Review Questions

The following questions may be assigned by your instructor as a way to assess your knowledge of this chapter. Your instructor has the answers to the review questions.

1. Some commands listed in the menu (drop-downs) have a keyboard shortcut listed next to them. (T/F)

2. An item in the menu with a check next to it indicates it is active. (T/F)

3. A toolbar cannot be moved. (T/F)

4. The right-click menu is always the same list of options. (T/F)

5. Some tools in the menu are grayed out, meaning they are inactive, when not relevant to the current drawing for selection. (T/F)

6. The icon with the hand picture ( ) allows you to _____ within a model.

7. The toolbar that you see when SketchUp is first started is the only toolbar available within the program.

8. What is the area, where you enter lengths, to the far right of the *Status Bar* called?

_____ _____ _____

Chapter 3
Open, Save and Close

Opening **SketchUp** is just like opening most any other program. You can either locate the file, i.e. the 3D model, using *Windows Explorer* (aka, *My Computer* or just *Computer*) and then double-click on the file, or you can open SketchUp and then create a new file or open one previously created.

If SketchUp is properly installed on your computer, you can launch the program from the Windows *Start* menu. To do this, make the following clicks within the *Start* menu (Figure 3.1):

Start → All Programs → SketchUp 2013 → SketchUp

Or double-click the **SketchUp 2013** icon from your desktop.

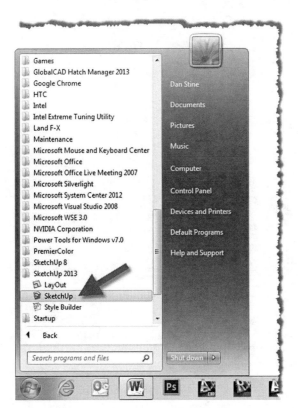

This may vary slightly on your computer depending on the version of Windows you are using (or if you are using a Mac); see your instructor or system administrator if you need help. It is possible to have more than one version of SketchUp installed on a computer. Make sure you are using version 2013 to gain access to all the new features and to ensure your screen matches the images in this book! The previous version is called SketchUp 8.

FIGURE 3.1 Welcome to SketchUp Interface

Start a New SketchUp Model

By default, SketchUp will open in the *Welcome to SketchUp* dialog as shown (Figure 3.2). Here you have access to more ways to help yourself learn SketchUp and to the various templates provided. Make sure you select the correct template before clicking the **Start using SketchUp** button.

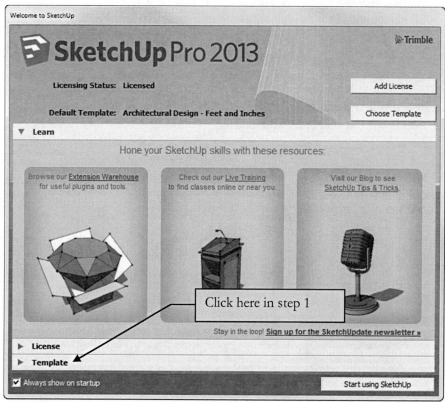

FIGURE 3.2 Welcome to SketchUp Interface

> *TIP: Experienced SketchUp users will uncheck the "Always show on startup" option in the lower left so they can get right to work. You should NOT uncheck this option until you consider yourself somewhat proficient in the program. FYI: If running SketchUp Pro in trial mode, you will not see this option.*

Here you will learn how to start a new SketchUp model.

1. Click the arrow next to the word *Template* (pointed out in the Figure 3.2).

SketchUp provides several templates. You will be selecting the one set up with the architect and interior designer in mind: *Architectural Design – Feet and Inches*.

2. Click to select **Architectural Design – Feet and Inches** from the list of available templates (as shown in the image below).

    FYI: *This will be the default template selected the next time SketchUp is opened.*

3. Click the **Start using SketchUp** button in the lower right (Figure 3.3).

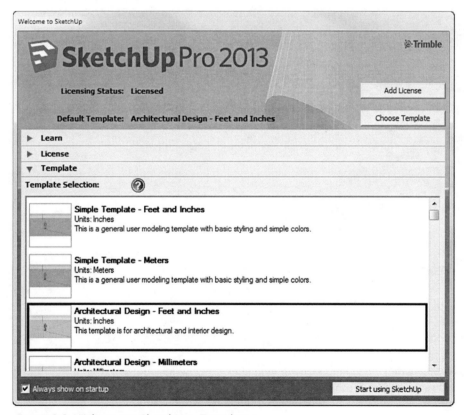

FIGURE 3.3 Welcome to SketchUp - Templates

You are now in a new SketchUp file! Notice the red, green and blue axis lines, the person pre-loaded (which is a great scale reference), and the implied ground which extends to the horizon. At this point you are in an unnamed file. The first time you click **Save**, you will be prompted to select a location and provide a file name; make sure you pay close attention to where you save the file and what you call it!

Open an Existing SketchUp Model

Now that you know how to open SketchUp and create a new file in which to model, you will open an existing SketchUp file. You will select a sample file provided on the book's webpage.

4. Download the "Required Files" from www.SDCpublications.com (Figure 3.4).

FIGURE 3.4 Download required files for this book

5. In SketchUp (follow steps 1-3 above), Select **File → Open**.

    TIP: *Pressing **Ctrl + O** will also get you to the Open dialog.*

6. In the *Open* dialog, browse to your hard drive where you save the files downloaded in step 4 and select **Office Building.skp**.

7. Click **Open**; if you are prompted to save the current model, choose **No**.

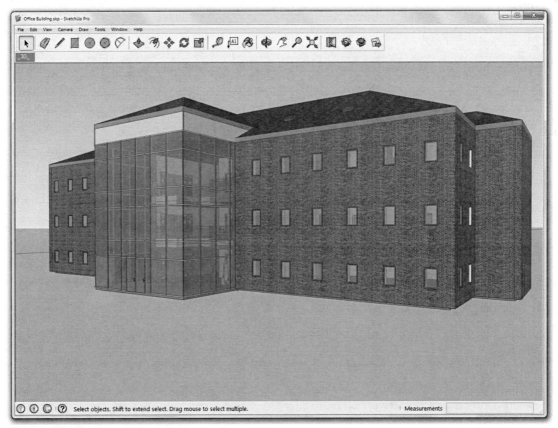

FIGURE 3.5 File opened

You are now in the *Office Building* file (Figure 3.5). Because you did not do any work in the new file you had just created, SketchUp discards that file in favor of the file you are opening. However, if you made any changes, SketchUp would have prompted you to save before closing the file. In SketchUp you can only have one file open at a time, but it is possible to have multiple sessions of SketchUp open – each with a different file.

TIP: To open multiple copies of SketchUp, simply double-click on the desktop icon or the SketchUp icon via Start menu.

Closing a SketchUp File

Because you can only have one file open at a time, and one file must be open, the only way to "close" a file is to open another file or exit SketchUp altogether.

The previous section discussed opening a file, and exiting SketchUp will be coming up soon.

If you have not saved your file yet, you will be prompted to do so before SketchUp closes. **Do not save at this time.**

Saving a SketchUp Project

NOTE: At this time we will not actually save a project.

To save a project file, simply select **Save** from the **File** menu. You can also press **Ctrl + S** on the keyboard.

When the *Standard* toolbar is open, you can also click the **Save** icon.

You should get in the habit of saving often to avoid losing work due to a power outage or program crash. The program automatically creates a backup file every time you save; that is, the current SKP file is renamed to SKB. So the SKP file will have the most current model and the SKB will be the state of the model the last time you saved. The backup and auto-save options can be set via **Window** (menu) ➔ **Preferences** (Figure 3.6).

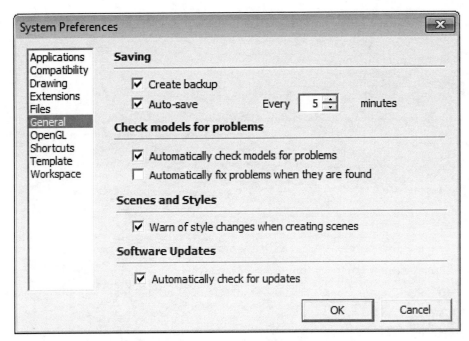

FIGURE 3.6 System Preferences Dialog

Auto-save files are saved in the same folder as the file. If SketchUp crashes, the file can be used to recover what would otherwise be lost work. When SketchUp closes properly, the auto-save file is deleted and thus cannot be accessed.

Closing the SketchUp Program

Finally, from the **File** menu select **Exit**. This will close the current file and shut down SketchUp. Again, you will be prompted to save, if needed, before SketchUp closes. **Do not save at this time**.

You can also click the red "X" in the upper right corner of the SketchUp application window.

Project Example: Reception Desk.

Image courtesy of LHB; www.LHBcorp.com

Chapter 3 Review Questions

The following questions may be assigned by your instructor as a way to assess your knowledge of this chapter. Your instructor has the answers to the review questions.

1. The files required for the tutorials in this book must be downloaded from the following website: _____

2. You can set the time interval for aut0-saves. (T/F)

3. SketchUp can have more than one model open during the same session. (T/F)

4. If the model has unsaved changes, SketchUp will prompt you to save before allowing you to close the model. (T/F)

5. To close a SketchUp model you must either open another file or close the application. (T/F)

6. Pressing **Ctrl + O** displays the _____ dialog box.

7. The primary SketchUp file extension (not the backup) is: _____.

8. *Auto-save* files are saved in the same folder as the SketchUp model. (T/F)

Chapter 4
Viewing SketchUp Models

Learning to get around in a SketchUp model is essential to accurate and efficient design and visualization. We will review a few tools and techniques now so you are ready to use them with the first design exercise.

You will select a sample file from the files downloaded previously in Chapter 3.

1. Open SketchUp and then select **File → Open**.

2. Browse to the folder where you saved the files downloaded from the publisher's website: www.SDCpublications.com. If you did not download the files yet, refer back to Figure 3.4 for more information, if necessary.

3. Select the file **Office Building.skp** and click **Open** (Figure 4.1).

You should see a view of the SketchUp model similar to that shown in Figure 4.1.

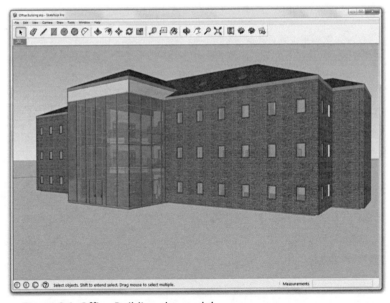

FIGURE 4.1 Office Building.skp model

Using Zoom and Pan Tools

You can access the navigation tools from the *Getting Started* toolbar – shown in the image below. The tools are, from left to right: *Orbit, Pan, Zoom* and *Zoom Extents*.

These tools do the following:

- Orbit: Fly the camera view around the model
- Pan: Pan the camera view vertically and/or horizontally
- Zoom: Zooms in or out – centered on current view
- Zoom Extents: Zooms view so everything in the model is visible

You will now have an opportunity to try each of these tools in the sample model.

Orbit

4. Select the **Orbit** icon from the *Getting Started* toolbar.   *Keyboard Shortcut*: **O**

5. Drag your cursor across the screen from right to left – holding down the left mouse button. Stop when your view of the building looks similar to Figure 4.2.

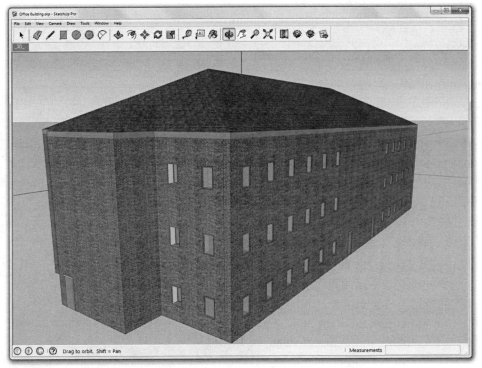

FIGURE 4.2 Using the Orbit tool

Spend a little time using the *Orbit* tool, looking at the model from the top, bottom and each side.

6. Click the **Select** icon to cancel the current tool and get back into the default mode of being able to select things in the model.

FIGURE 4.3 Scene tab

7. Once you are done experimenting with the *Orbit* tool, you can quickly get back to your original view by clicking the "_3D_" **scene tab** (Figure 4.3).

Additional *scene tabs* can be added, saving views of different parts of the building – both interior and exterior. The scene tabs can also be used to define the outline of an animation, where SketchUp smoothly transitions from location to location. This animation can also be exported and shared with others or used in a presentation. See Chapter 12 for more on scenes and animations. Also, try opening the sample project, **LHB Project Sample – Hybrid Medical Animation.skp**, and click on the two tabs to switch between in the interior and exterior of the model (see image to right).

Provided sample project with scene tabs

Pan

The **Pan** tool allows you to reposition the camera left/right or up/down relative to the current view direction. This is helpful if a portion of the building extends off the screen and you want to see it, but you do not want to change the angle of the view (i.e. see more of the side rather than the front) – as the *Orbit* tool would do. This tool will be particularly useful when composing interior views. Next you will test-drive the *Pan* tool.

8. If your view is not reset, do so now per the previous step.

9. Click the **Pan** icon from the *Getting Started* toolbar.   *Keyboard Shortcut:* **H**

Notice how the cursor has changed to a hand symbol to let you know the *Pan* tool is active.

10. Drag the cursor from right to left until the view looks similar to Figure 4.4).

As you can see, the camera moved, which is similar to you walking by a building. As you walk by, you see a little more of what is around the corner than when you were right next to the front. Later you will learn how to toggle between **Perspective** and **Parallel Projection**, the first being more realistic with vanishing lines, the latter is like a flat 2D drawing. When the view is in *Parallel Projection* mode, you do not see more or less of anything – the view stays the same, it is just being moved around on the screen.

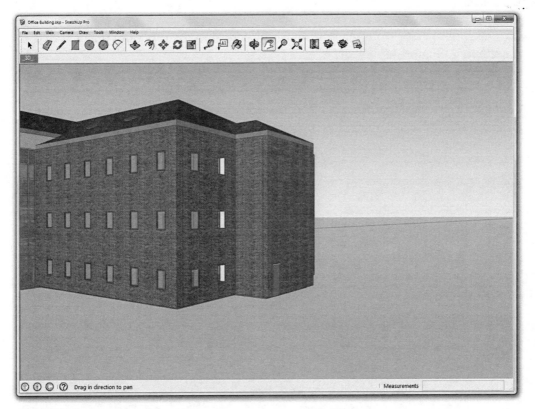

FIGURE 4.4 Using the Pan tool

Zoom

The *Zoom* tool basically does what the name implies… it zooms in and out of your model. Keep in mind it is not changing the size of anything. This feature is not the same as zooming in and out on your camcorder. With the camcorder analogy, in SketchUp you would actually be walking closer to the building when zooming rather than simply magnifying an area. In fact, you can zoom in so far you actually enter the building. Don't forget, you can click the *scene tab* to quickly restore your view if things get messed up.

11. Select the **Zoom** icon from the *Getting Started* toolbar.   *Keyboard Shortcut:* **Z**

Notice the cursor changes to a magnifying glass symbol to let you know you are in the *Zoom* command. This will be active until you press the **Esc** key, click to start another command or click the **Select** icon.

12. Drag your cursor from the bottom of the screen to the top, until the view looks similar to Figure 4.5.

The view is zoomed relative to the center of the drawing window. You will learn a better way to zoom coming up, which allows better control of where you zoom. Notice how the high quality textures (i.e. building materials) appear more realistic the closer you get.

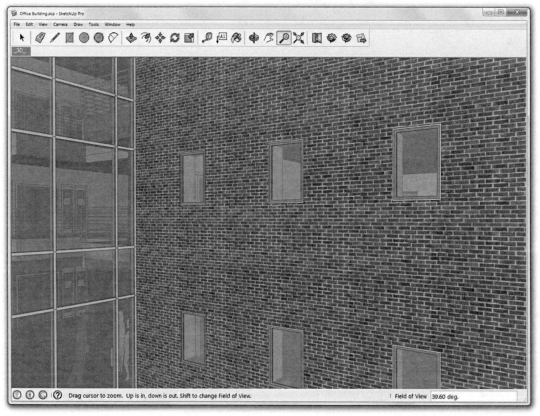

FIGURE 4.5 Using the Zoom tool

Sometimes you need to make several dragging motions with your mouse to zoom in far enough (because there is not enough room on the desk or your arm simply will not reach). To do this, click and drag as far as you can, and then release the mouse button, move your mouse back, and then repeat the process (i.e. click and drag).

Dragging the mouse in the opposite direction zooms out.

13. Try zooming out, using the **Zoom** tool, dragging from top to bottom.

14. When finished testing the *Zoom* tool, click the **scene tab** to reset the view.

> **FYI:** *Holding down the* Shift *key while zooming changes the field of view degrees. This is similar to changing the lens on a camera. A larger angle gives you a wide-angle view, allowing you to see more in a smaller space. However the view can be more distorted as the angle is increased*

Zoom Extents

The **Zoom Extents** tool is a quick way to make sure you are seeing everything in the model from your current vantage point. You simply click the icon and SketchUp does the rest. This can be tricky if something is floating way out in space because using *Zoom Extents* will show a stray line and the rest of your model on the same screen – which means your model might be a tiny dot on the screen somewhere.

15. Try the **Zoom Extents** tool:

    a.   Zoom in on the building (similar to Figure 4.5).

    b.   Click the **Zoom Extents** icon.   *Keyboard Shortcut*: Shift + **Z**

In this example this would not be any faster than clicking the *scene tab*. However, there will not always be a corresponding *scene tab* for every angle from which you will be looking at your model. So the *Zoom Extents* tool is very useful.

Using the Scroll Wheel on the Mouse

The scroll wheel on the mouse is a must for those using SketchUp. In SketchUp you can *Zoom* and *Orbit* without even clicking the *Zoom* or *Orbit* icons. You simply **scroll the wheel to zoom** and **hold the wheel button down to orbit**. This can be done while in another command (e.g., while sketching lines). Another nice feature is that the drawing zooms into the area near your cursor, rather than zooming only at the center of the drawing window. Give this a try before moving on. Once you get the hang of it you will not want to use the icons. The only thing you cannot do is *Zoom Extents* so everything is visible on the screen.

To use the ***Pan*** feature (aka, *hand* tool), simply hold down the **Shift** key while pressing the center wheel button.

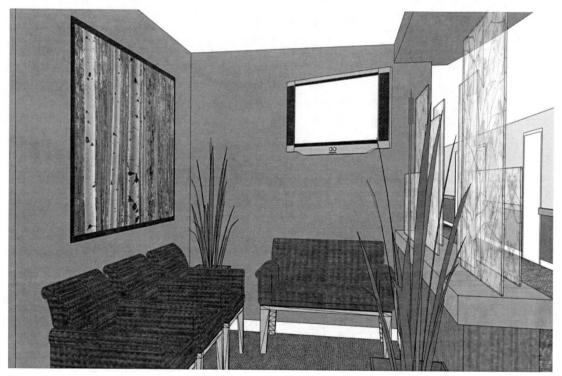

Project Example: Waiting Area.   *Image courtesy of LHB; www.LHBcorp.com*

Chapter 4 Review Questions

The following questions may be assigned by your instructor as a way to assess your knowledge of this chapter. Your instructor has the answers to the review questions.

1. Scene tabs can be used to define the outline of an animation. (T/F)

2. Spinning the wheel on your mouse allows you to quickly zoom in or out. (T/F)

3. Holding down the Shift key while zooming changes the field of view degrees. (T/F)

4. SketchUp will zoom into the area directly below your cursor. (T/F)

5. What is the keyboard shortcut for Pan: _____

6. Pressing the wheel button on your mouse allows you to quickly pan in your model. (T/F)

7. Pressing Shift + Z initiates the _____ _____ command.

8. SketchUp is not capable of the more realistic "perspective" mode og viewing a model. (T/F)

Chapter 5
Help System

Using the *Help* system is often required when you are having problems or trying to do something advanced. This section will present a basic overview of the *Help* system so you can find your way around when needed. It is important that you don't skip this section as it can help reduce your stress level when/if you run into problems.

1. Open SketchUp, if not already open. It does not matter if you are in a blank file or a sample file.

2. Select **Help → Knowledge Center**.

3. Your default internet browser opens and you are in SketchUp's *Help* system (Figure 5.1). The entire *Help* system is internet based, thus allowing Trimble the ability to make revisions and additions as needed.

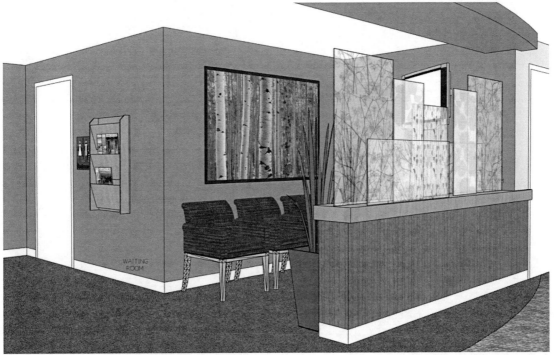

Project Example: Art Glass Feature Wall.   *Image courtesy of LHB; www.LHBcorp.com*

FIGURE 5.1 Help system interface

Searching for Answers

The various links found on the *Help* page speak for themselves. However, most of the time you can just type a command name and press **Enter** to get a refined list of options from which to choose. You will try that now.

4. Click within the search box and type **ORBIT**.

5. Press **Enter** or click the **Search Help** button to the right.

FIGURE 5.2 Entering a word to search for within Help

The image below shows the search results (Figure 5.3). Since this is a web based search, the results can change over time. Of course, you also need internet access for the search to work.

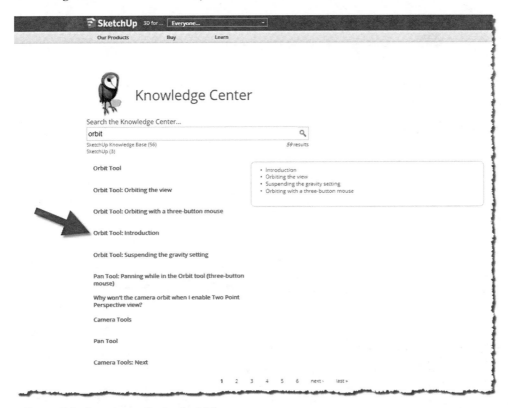

FIGURE 5.3 Search results for "orbit"

The most relevant item may not always be first. The example's first item relates more to the mouse rather than *Orbit*. The first place to start might be the introduction link, which is third in the list (Figure 5.3). Also notice the word(s) you typed is bold in the title and description.

6.   Click on the link titled **Orbit Tool: Introduction**.

Your result should look similar to Figure 5.4. Notice the various items: basic description, keyboard shortcut, tags.

FIGURE 5.4 Sample help item

7.   Try clicking on the links on the right (i.e., *Zoom Extents*, *Pan Tool* and *Rotate Tool*) to see the information that is shown (see arrow in image above).

If you are searching for a multi-word tool it is best to add quotation marks around the entire search text to narrow the search. For example, if you want to search for information on the *Zoom Extents* tool, you should enter:

- "zoom extents" rather than: zoom extents

SketchUp Help will show results with both words, but if quotation marks are not used, the results will be different.

Anytime you want to return to the initial help screen you can click the "SketchUp Owl" image-link in the upper left.

SketchUp Forum

In addition to the formal help system SketchUp as provided, they also host a user to user forum. A link to this can be found at the bottom of the Help screen. On the forum you may ask and answer questions. You can even just review other questions and answers as a way of learning more about the program and problems others have had. Sometimes you need to be careful with the answers provided. Just because someone relied does not mean they fully understand your problem, computer setup and project needs. Bad advice can make things much worse. However, most of the time the help provided is invaluable and can save tons of time.

SketchUp User's Guide

Another way to learn and do research on SketchUp is via the user's guide. This is an indexed list which makes it easy to find information on a specific topic (e.g., placing a camera). You might not even know the name of the command you want to use. In the user's guide you can look for it by process of elimination. This is a great way to stumble across information you were not even looking for – similar to randomly opening a book to a page and something catches your eye, so you start reading about it.

The user's guide can be found via a link on the initial help page. Click the plus symbols to expand (or close) a section (Figure 5.5).

Be sure to refer to the *Help* system anytime you get stuck to see if it can help you find the answer to your problem.

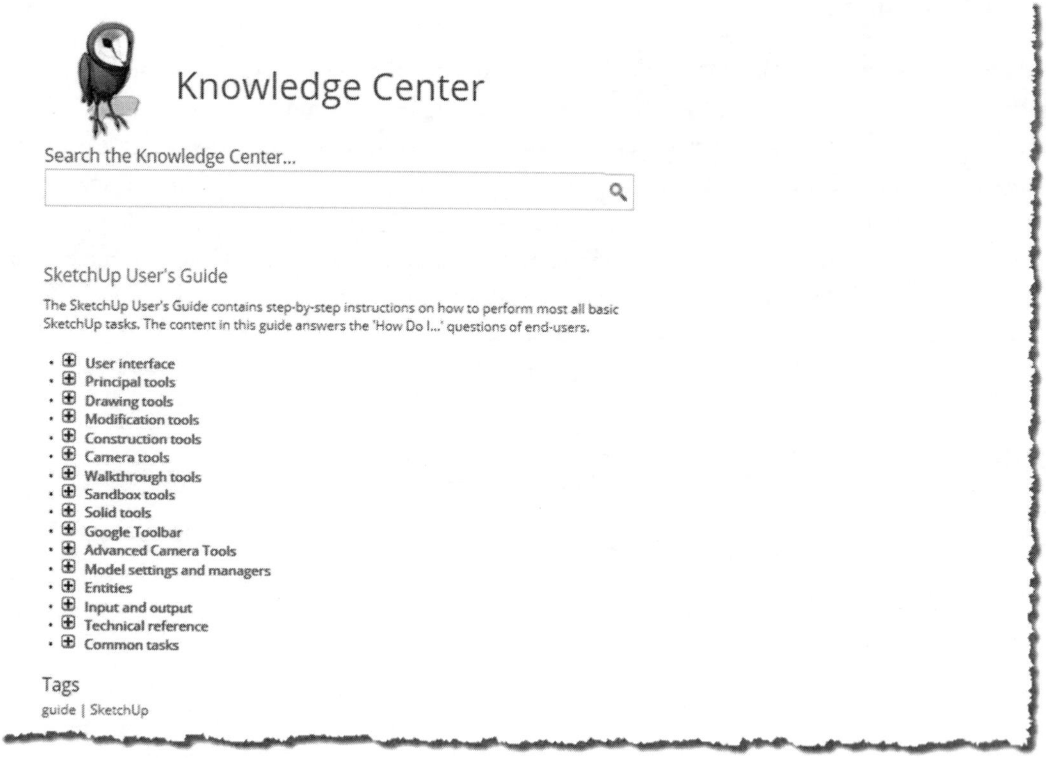

FIGURE 5.5 User's guide index

Project Example: Hallway Flooring.

Image courtesy of LHB; www.LHBcorp.com

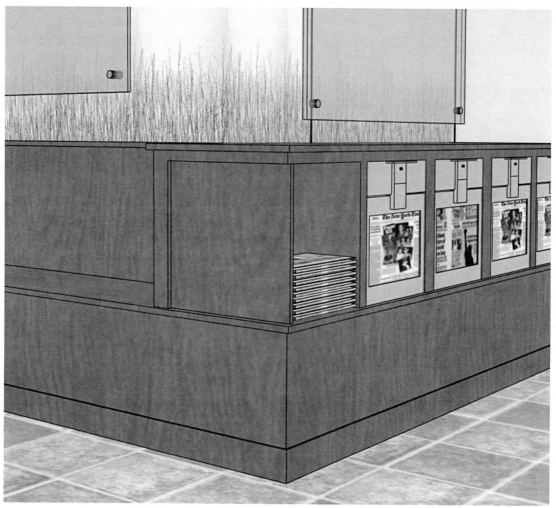

Project Example: Custom Casework.

Image courtesy of LHB; www.LHBcorp.com

Chapter 5 Review Questions

The following questions may be assigned by your instructor as a way to assess your knowledge of this chapter. Your instructor has the answers to the review questions.

1. You do not need an internet connection to use the Help system. (T/F)

2. You may access the SketchUp **User's Guide** via the Help System (aka Knowledge Center). (T/F)

3. When reviewing a topic in help, there are "related" topics listed on the side margin. (T/F)

4. SketchUp hosts an online forum where other users may answer your questions. (T/F)

5. The most relevant item will always be first when searching the help system. (T/F)

6. You cannot search the knowledge center. (T/F)

7. You access help (aka Knowledge Center) via the _____ menu.

8. Using the *Help* system is often required when you are having problems. (T/F)

Chapter 6
The Basic Entities

Given the amazing images one can create using SketchUp, it may be somewhat surprising that there are mainly eight types of entities which can be added to a model.

They are:

- Edges
- Surfaces
- Annotation
    o Dimensions
    o Text
    o 3D Text
- Components
- Groups
- Guide (reference line)

The next few pages will provide a brief overview of each of these entity types.

Edges

SketchUp is a face-based program, and all surfaces (i.e. faces) must be defined by an edge. This is the fundamental building block of a SketchUp model.

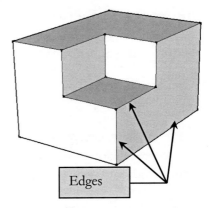

FIGURE 6.1 Edges pointed out

Edges are created with one of the *Draw* tools:

- Rectangle
- Line
- Circle
- Arc
- Polygon
- Freehand

Edges can be created with very specific length (or radius), or arbitrarily by clicking anywhere within the drawing window. It is easy to snap to one of the three planes (axes) while drawing lines. This makes it possible to draw 3D shapes from a single 3D view (more on this later).

A basic cube has twelve edges. Figure 6.1 has 21 edges and Figure 6.2 has two edges.

Even circles are made up of small edges. When one is being created, the *Value Control Box* (on the *Status Bar*) lists the number of sides that will be used to approximate the circle. This number can be increased to make larger circles smooth, or decreased to make smaller circles less complex (which can be a burden on model performance).

Later in this chapter you will get some practice drawing edges and editing them.

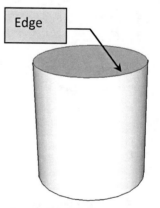

FIGURE 6.2 Edge pointed out

Edges can be modified with a number of tools. For example, an edge can be scaled, rotated, divided, copied, offset and erased. These tools are accessible from the *Tools* menu, toolbars (which may not be visible yet), right-click menus (when the edge is selected) and keyboard shortcuts.

Edges can also be placed on *Layers* in order to control visibility. A *Layer* can be turned off, making everything assigned to that *Layer* invisible.

Surfaces

A surface is the second most significant type of entity in SketchUp. You might be surprised to learn that no tool exists to create a surface! They are created automatically when the conditions are right.

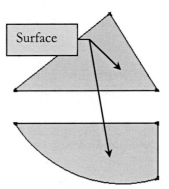

What are the conditions in which a surface is automatically created? The simple answer is that **a series of edges forms an enclosed area**. When the last edge is drawn which defines an enclosed area, a surface is created. This can be as few as three edges – forming a triangle.

The image to the right (Figure 6.3) shows two examples of three connected edges defining a *surface*. Note the edges can be a combination of straight and curved lines.

FIGURE 6.3
Surfaces defined by at least three edges

If an *edge* is erased, the *surface* is also erased, seeing as it no longer has a boundary.

In addition to *edges* forming a closed perimeter, there is another important requirement a new modeler needs to be aware of; that is, **all the lines forming the enclosure must be coplanar**.

Project Example   *Image courtesy of LHB; www.LHBcorp.com*

If you don't already know, the easiest way to describe coplanar is to think of all the edges as lines drawn on a flat piece of paper. As long as all the lines are in the same plane (i.e., on that flat piece of paper) a *surface* will be created (Figures 6.4 & 6.5).

Surfaces may have materials painted on them. They can also be placed on *Layers* in order to control visibility.

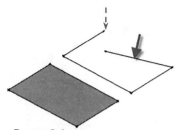

FIGURE 6.4
Coplanar on left, not on right

A surface can be deleted; simply select it and press the **Delete** key on the keyboard. The only way to get another surface is to draw a line directly on top of one of the existing *edges*. SketchUp will then create a surface and delete the extra line, as it does not allow two lines to exist directly on top of each other.

FIGURE 6.5 Another angle of Figure 6.4

Dimensions

Dimensions can be added to your SketchUp model. These are smart entities; they are not sketched lines and manually entered text. A *dimension* entity becomes a permanent part of the model, unlike the *Tape Measure* tool (which is used to list distances without drawing anything).

To place a dimension you simply pick three points; the first two are what you want to dimension and the third is the location of the dimension line and text. SketchUp automatically displays the correct length.

The dimensions are associative, relative to the first two points picked. The dimension will grow or shrink if the geometry is modified. However, if the geometry is deleted the dimension will remain (but is no longer associated to anything).

Dimensions can be tricky in that they may appear correctly and legibly from one angle (Figure 6.6a) but not another (Figure 6.6b). But the visibility of a *dimension* can be controlled by *Layers* or by *Scene* (more on what scenes are later).

To adjust various settings related to how dimensions are created, go to **Window → Model info** and then click **Dimensions** in the list on the left.

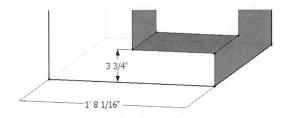

FIGURE 6.6A   Dimensions added while viewing the model from this angle

Text

SketchUp has a tool which allows you to add notes with leaders (a leader is a line from the text, pointing at something). To place a *Text* entity, you make two clicks and then type (or accept the default value). Default value? If you point to a *surface*, SketchUp will automatically list the area of that surface. If an *edge* is pointed to, SketchUp will list its length. An example of each can be seen in the image on the next page (Figure 6.7).

If you don't want a leader, simply click in empty space and you can just type text. The text "Option A" is an example of text without a leader (Figure 6.7).

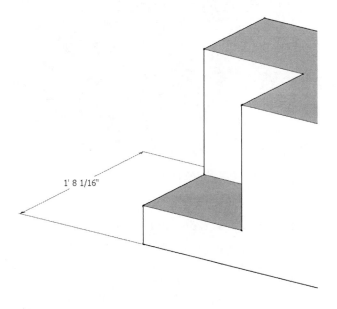

FIGURE 6.6B   Only one dimension still visible when view angle is changed using orbit

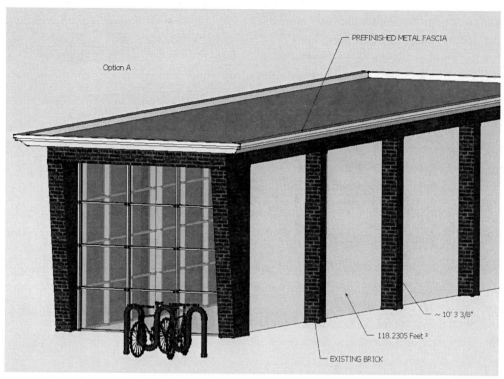

FIGURE 6.7 Notes added using the Text tool

Similar to dimensions, text entities remain visible when the vantage point is changed, using *Orbit* for example. As you can see in the Figure 6.8, this can get a bit messy. Also like *dimensions*, *Text* visibility can also be controlled with *Layers* and *Scenes*.

For the most part, notes and dimensions are left until the end of the modeling or added outside of SketchUp – in LayOut or a CAD program such as AutoCAD or Revit.

When text is right-clicked on, a menu pops up which allows you to change the arrow type and leader (Figure 6.9). These options, plus the ability to change the font, are available via the ***Entity Info*** dialog. This can be turned on from the *Window* pull-down menu.

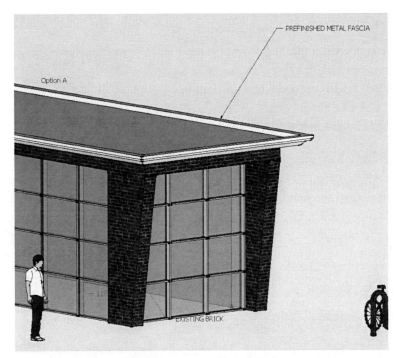

FIGURE 6.8 Notes still visible when model rotated using orbit

The default settings for text can be changed via **Window → Model Info**, and then selecting **Text** from the list on the left.

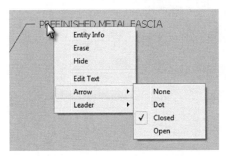

FIGURE 6.9 Right-click options for text

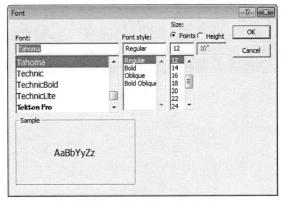

FIGURE 6.10 Changing font for text entity

3D Text

The *Text* tool, just covered, is meant for notes and comments about the model. *3D Text* is meant to be part of the model. This tool is used to model text on signs or letters on the face of a building. Unlike notes created with the *Text* tool, *3D Text* stays right where you put it.

Placing *3D Text* is easy. Select the tool, and the *Place 3D Text* dialog appears (Figure 6.11). Enter your text and select the options desired for font style, height and thickness (i.e., extruded). Click **OK** and then pick a location on a face to place it.

Once the 3D Text is placed, the *Paint Bucket* tool can be used to apply a material.

FIGURE 6.11 Adding 3D Text

FIGURE 6.12 3D Text placed

Once the text is created, it becomes a *Component* that cannot be easily edited (in terms of typing new words).

It is possible to see the properties for *3D Text*, or anything else selected, using the ***Entity Info*** dialog (Figure 6.13). This can be turned on from the *Window* pull-down menu. The information presented varies depending on what is selected. This dialog can remain open while modeling.

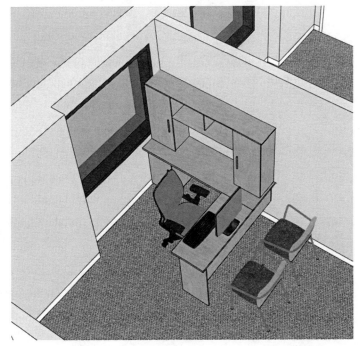

Entity Info

Solid Component (1 in model)

Layer: Layer0

Name:

Definition Name: BIKE SHOP

Volume: 215.9327 Inches

☐ Hidden ☑ Cast Shadows
☐ Locked ☑ Receive Shadows

FIGURE 6.13 Entity Info dialog

Components

In SketchUp one can think of *Components* being something like clipart in a word processing program – but clipart on steroids! They are pre-built models which can be reused in your SketchUp model. Some *Components* are flat 2D models while others are complex 3D models. The simple, flat *Components* reduce the resources required of your computer, making it easier to smoothly orbit and inspect your model. For example, many of the trees which designers use in SketchUp are 2D due to the large number typically needed. If 3D trees were used, the file would be large and unmanageable. The 2D *Components* can be setup so that they always face you – plus they cast shadows (see Figures 6.14 and 6.15).

Image courtesy of LHB; www.LHBcorp.com

FIGURE 6.14 2D vs. 3D components; two items are 2D and two are 3D.

FIGURE 6.15 Rotated view of previous image

Right-clicking on a *Component* allows you to edit it, explode it (reduce it down to individual entities) and add parameters and parametrics using the advanced **Dynamic Components** functionality.

Editing a *Component* causes all instances of that *Component*, in your model, to instantly update. You will see an example of this in the next section.

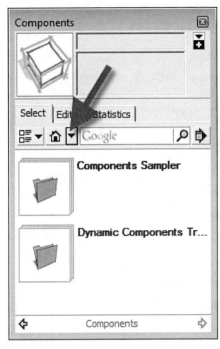

FIGURE 6.16 Components dialog

One of the truly great things about using SketchUp is the amount of content the designer has access to. Google hosts a site called ***Trimble 3D Warehouse*** which has thousands of *Components* ready for the taking.

Some of the content found on *Trimble 3D Warehouse* is provided directly by Trimble, while other content comes from manufacturers of products (who hope you will ultimately buy or specify their products) or from end users like you.

Of course, *users beware* on anything one downloads and uses in their design. As a design professional (or would-be, someday, design professional) you are responsible for code and performance compliance. So you cannot just assume the toilet or the door you downloaded is the correct size. You need to double-check it with the manufacturer's data sheets. Now, if the content was created by the manufacturers, it is highly probable that it is the correct size.

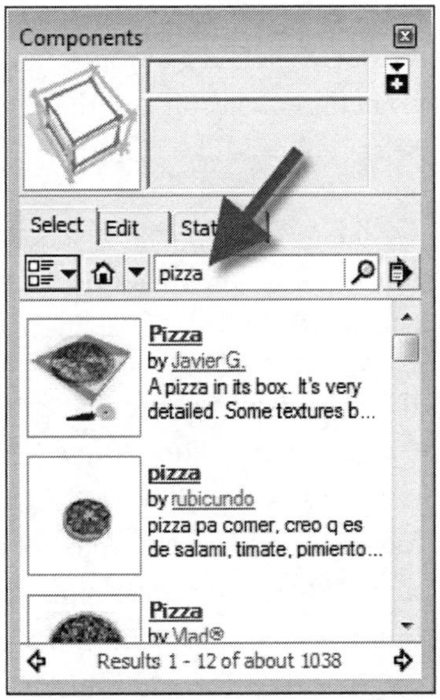

FIGURE 6.17 Components search

The *Components* dialog (Figure 6.16) is the easiest way to add *Components* to your model. This can be accessed from the *Window* menu. The down arrow highlighted reveals a menu which provides shortcuts to groups of content, such as *Architecture*, *People*, *Playground*, etc.

It is also possible to search for *Components*. You may be surprised at what you can find. Figure 6.17 shows some of the results when searching for "**pizza**"! Notice the author of the component is listed directly under the name.

Try a few searches to see what you can find – maybe try goat, newspaper, or snowboard.

See the next section for more on *Components*.

Groups

A *Group* is similar to a *Component* in that you can select one part of it and the entire representation is selected (selecting potentially hundreds of entities with a single pick). However, that is about all that is the same between them.

Groups are meant for one-off type items; that is, a unique reception desk, a built-in entertainment center, etc. A *Component* is used when your model will contain many instances of an object.

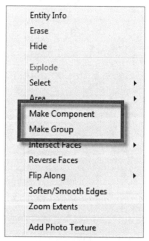

Both *Groups* and *Components* are easy to create. You simply model something, select it and then right-click (on it). At this point you can select either **Make Group** or **Make Component** (Figure 6.18).

Both *Groups* and *Components* can be copied around the model (using the *Move* tool and holding down Ctrl). They both can also be edited, by right-clicking and selecting "edit" from the pop-up menu.

It is important to note that editing a *Group* only changes the specific *Group* you are editing. But editing a *Component* instantly causes all instances of that *Component* to update (see Figures 6.20 and 6.21). This means SketchUp duplicates all information required to define each copy of a *Group*. A single definition is all that is needed for multiple instances of a *Component*. Of course, this means a file with many copies of a *Group* will be larger than one with many copies of a *Component*.

FIGURE 6.18 Right-click menu

The main thing to keep in mind is that *Groups* are quick and require minimal decisions. *Components* can be much more sophisticated and take a lot of time setting up (creating parameters and parametric relationships, and adding formulas).

When you right-click and select **Make Group**, SketchUp just makes it without asking any questions. It can be selected and named via the *Entity Info* dialog if you wish.

When creating a *Component*, the *Create Component* dialog appears (Figure 6.19). Notice the various options:

- Glue to – does the tree stick to the ground or float in the air?

- Always face the camera – this is ideal for flat two-dimensional items.

- Replace selection – turn the current selection into one of the *Components* you are creating.

FIGURE 6.19 Right-click menu

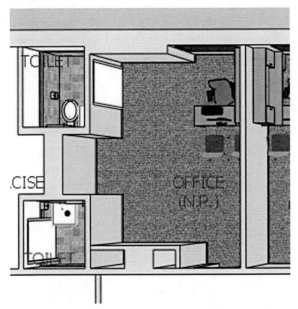

Image courtesy of LHB; www.LHBcorp.com

The images on the next two pages compare what happens when a *Component* is edited versus a *Group*. Notice all instances of the *Component* are updated, whereas only the selected *Group* being edited is updated (even though the other *Groups* are copies of the one being edited).

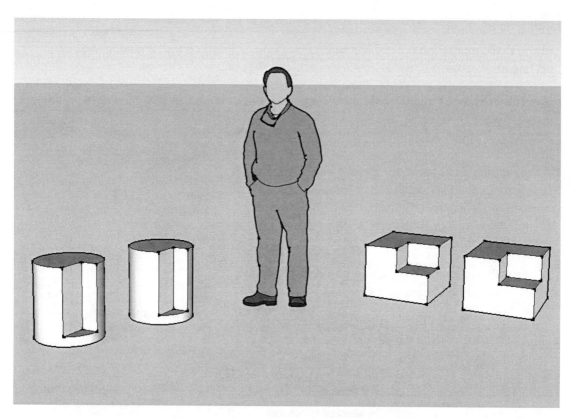

FIGURE 6.20 Components vs. Groups – Components on the left; Groups on the right

FIGURE 6.21 Components vs. Groups – all Components update; only selected Group updates

Guides

Guide lines (or construction lines) are useful for new users and for a general design reference grid. The image below (Figure 6.22) shows the main *Axes* and a few *Guides* at 5'-0" intervals. These lines are parallel to the main *Axes* and are infinite in length. Note how they converge at the horizon line.

The ***Tape Measure*** tool is used to create *Guide* lines. Follow these simple steps to create one:

- Start the *Tape Measure* tool.
- Click on the *Edge* of any shape or *Axes*.

> **FYI:** *Clicking on an endpoint creates a Guide Point.*

- Drag the cursor perpendicular to where you want the *Guide*.
- Release the mouse to locate the *Guide*.
- Type in a length to (retroactively) adjust the *Guide* location.

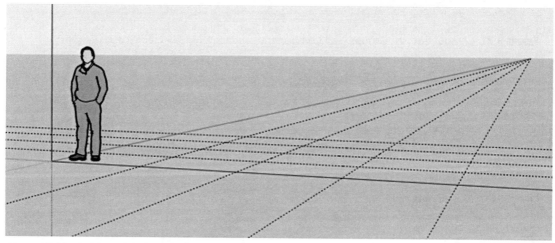

FIGURE 6.22 Guides added at 5'-0" intervals

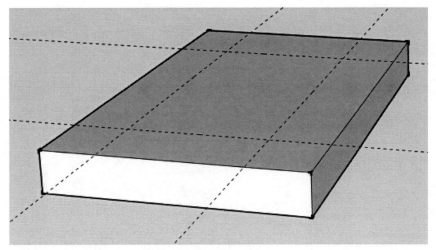

FIGURE 6.23 Guides added on top of a surface

FIGURE 6.24 Guides visibility

Guides can be selected and deleted. They can also be relocated with the *Move* tool. They can be rotated with the *Rotate* tool as well.

You can quickly hide the *Guides* via the *View* menu (Figure 6.24). Notice the *Axes* can also be toggled off/on here as well.

Guides can also be placed on a *Layer* and hidden. This would allow you to hide some *Guides* while leaving others visible. Simply create a *Layer* using the *Layer* dialog (*Window → Layers*). Then select the *Guide(s)* and switch them to another *Layer* via the *Entity Info* dialog (see page 47).

If you can see *Guides* on the screen, they will print. You need to hide them before printing if you do not want them to print.

Project Example: Reception Desk.   *Image courtesy of LHB; www.LHBcorp.com*

Chapter 6 Review Questions

The following questions may be assigned by your instructor as a way to assess your knowledge of this chapter. Your instructor has the answers to the review questions.

1.   A surface is automatically created when a series of edges forms an enclosed area.. (T/F)

2.   Text and dimension visibility can be controller using layers. (T/F)

3.   Changing one Component will change all instances of that component in the same model. (T/F)

4.   There are 8 basic types of entities in SketchUp. (T/F)

5.   To place a dimension you pick _____ points.

6.   Guides are created from within the _____ _____.

7.   SketchUp cannot automatically create 3D text; you have to model each letter. (T/F)

8.   Once created, a surface cannot be deleted without deleted an edge. (T/F)

Chapter 7
Beginning with the Basics

In this section you will practice sketching basic 2D lines and shapes to get the hang of using a few of the draw and modify tools, as well as specifying specific dimensions. In the next section you will circle back and see how easy it is to turn these 2D sketches into 3D drawings. Normally you would do the 2D line work and then immediately turn it into a 3D model. But we are breaking the process down and focusing on each part separately.

Setting up the Model

The first thing you need to do is start a new model and make a few adjustments. You will complete these steps for each drawing in this section, unless noted otherwise (UNO).

1. Start a new SketchUp model using the **Architectural Design – Feet and Inches** template.

2. Select the person *Component*, and press the **Delete** key, on the keyboard.

To break things down into the simplest terms, you will change to a non-perspective plan (or top) view.

3. From the *Camera* menu, select **Parallel Projection** (Figure 7.1).

4. Also from the *Camera* menu, select: **Standard Views → Top**.

You can also go to **View → Toolbar → Views** to turn on a toolbar which provides quick access to the standard views (top, front, iso, etc.).

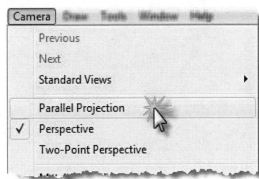

FIGURE 7.1 Parallel Projection mode

You are now looking at a plan view; notice the word "Top" in the upper left (Figure 7.2). This view is similar to what you would see on a printed out floor plan (aka blueprints or construction documents). Use caution not to press and drag your center wheel button as this action will activate the *Orbit* tool and throw you out of *Top* view; you would still be in "parallel" mode however. If you accidentally do this, simply select "top" again from the *Camera* menu. Selecting *Undo* does not help in this case.

Notice how the *Axes* are centered on the screen.

You can adjust which part of the model you are looking at using the **Pan** tool. When finished panning, click the *Select* icon. Do not pan at this time.

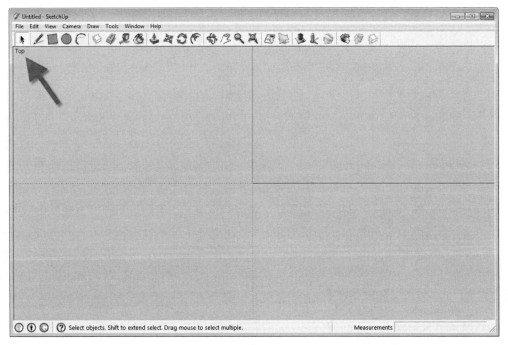

FIGURE 7.2 Top view in Parallel Projection mode

file name: **Bookcase**

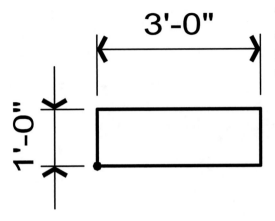

This is a simple rectangle that represents the size of a bookcase. The black dot represents the starting point, which should align with the intersection of the axes. **Do not draw the black dot.**

5. Select the ***Rectangle*** tool.

6. **For your first point,** click the intersection of the *Axes* (Figure 7.3).

Be sure your cursor snaps to the *Origin*; you will see a yellow circle and a tooltip appear.

7. **Select your second point** approximately as shown in Figure 7.3.

    a. You can keep an eye on the dimension box in the lower right, but do not worry about getting the number exact as that will be done in the next step.

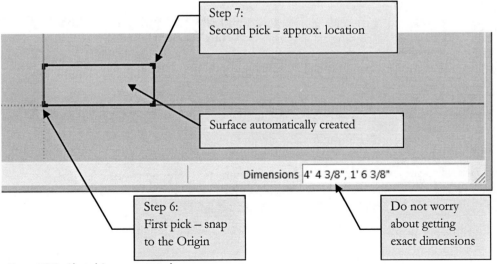

FIGURE 7.3 Sketching a rectangle

8. After clicking the second point (Step 7) and before doing anything else, simply type **3',1'** and then press **Enter**.

    a. You do not need to click in the *Dimensions* box; just start typing.

9. **Save** your file as **Bookcase.skp**.

Notice the surface which was automatically created once an enclosed area was defined. You are done with this file for now. You will come back to it later and turn it into a 3D bookcase.

file name: **Coffee Table**

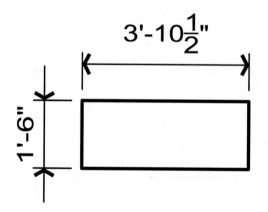

Next you will start a new file, using the steps previously covered (Setting up the Model; Steps 1-4). The previously created file can be closed and set aside for use in the next chapter.

The *Coffee Table* drawing will introduce you to entering fractional values.

10. Start a new model, following Steps 1-4.

It would be fairly easy to use the *Rectangle* tool again to draw this item, however you will use the basic *Line* tool so you can see how it works and get practice entering specific lengths.

11. Select the **Line** tool from the toolbar.

12. Snap to the *Origin* (the intersection of the red and green axes).

13. Begin moving your cursor to the right (Figure 7.4):

    a.   Ensure your cursor is "snapped" to the horizontal.

    b.   When horizontal, you should see the "**On Red Axis**" tooltip.

    c.   Once you are pointing in the correct direction and snapped to the horizontal, you may type in a length (see the next step for this).

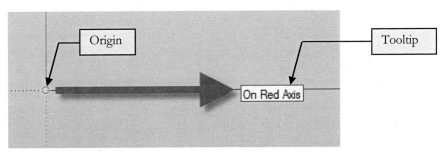

FIGURE 7.4 Sketching a line

14. Without moving the mouse, type **3′10.5** and then press **Enter.**

You always have to enter a foot symbol if feet are needed, however the inch symbol never needs to be typed as it is assumed when nothing is specified.

The *Line* tool will remain active until you pick the *Select* icon or another tool. Next you will draw one of the vertical lines.

15. While the **_Line_** tool is still active (Figure 7.5):

    a. Start moving your cursor straight up.

    b. Ensure the "**On Green Axis**" tooltip is showing, meaning vertical.

    c. Type **1′6** and then press **Enter**.

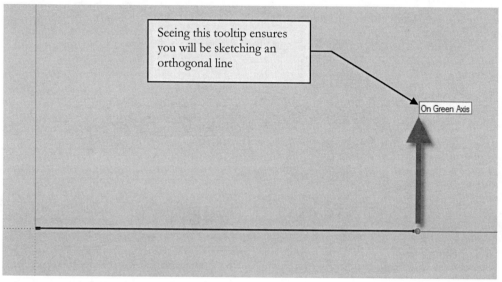

Seeing this tooltip ensures you will be sketching an orthogonal line

On Green Axis

FIGURE 7.5 Sketching another line

16. Using one of the alternative methods of entering 3′ 10½″ (see below), sketch the top horizontal line, from right to left.

<u>Entering fractions</u>: the 3'-10½" can be entered several ways.

○	3' 10.5	*Notice there is a space between the feet and inches.*
○	3'10.5	*Notice space can also be left out.*
○	3' 10 1/2	*Note the two spaces separating feet, inches and fractions.*
○	3'10 1/2	*The second space is always required.*
○	0' 46.5	*This is all in inches; that is, 3'-10½" = 46.5".*
○	46.5	*SketchUp assumes inches if nothing is specified.*

TIP: *Even when you are in a model using imperial units, you can type a metric value and SketchUp will automatically convert it. For example; typing **150mm** draws a **5 7/8"** line.*

To draw the last line you could type in the value but you can more quickly snap to the endpoint of the first line drawn. This will complete the rectangular shape.

17. Snap to the **Endpoint** of the first line drawn (Figure 7.6).

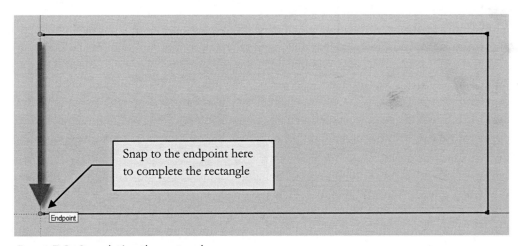

Snap to the endpoint here
to complete the rectangle

Endpoint

FIGURE 7.6 Completing the rectangle

Once a closed perimeter is defined as your rectangle, a surface is automatically created. You should use the ***Tape Measure*** tool too occasionally to double-check your lengths. Simply select the *Tape Measure* tool from the toolbar and then pick two points in the model. The *Value Control Box* in the lower right of the application lists the measurement. Give it a try!

18. **Save** your file as **Coffee Table.skp**.

file name: **Small Desk**

Next, you will start another new file and create this rectangular shape.

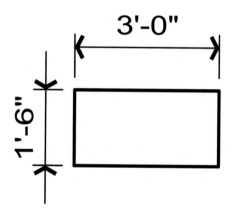

19. Create a new model (per Steps 1-4) and create this small desk using either of the methods just covered.

20. **Save** your file as **Small Desk.skp**.

Don't worry; things will get more challenging. These steps are laying the groundwork for all of the 3D modeling you will be doing! So make sure you take the time to understand this material.

file name: **Night Table**

Obviously, you could draw this quickly per the previous examples. However, you will take a look at copying a *file* and then modifying an existing model.

You will use the *Move* tool to stretch the 3'-0" wide desk down to a 1'-6" wide night table.

21. With the *Small Desk* SketchUp model still open, select **File → Save As**.

22. Type **Night Table** for the *File name* and click **Save** (Figure 7.7).

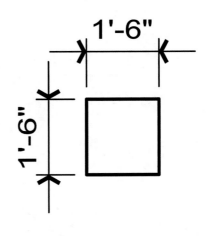

You are now in a new file named *Night Table.skp*, and are ready to manipulate the file. The original "small desk" file is now closed and will not be affected.

You will use the *Move* tool to change the location of one of the vertical lines, which will cause the two horizontal lines to stretch with it. SketchUp's *Lines* automatically have a parametric relationship to adjacent lines when their endpoints touch each other.

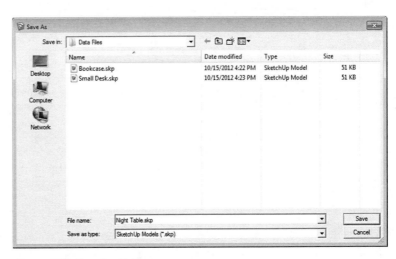

FIGURE 7.7 SaveAs dialog

23. Click the vertical line on the <u>right</u> and then select the **Move** icon on the toolbar (Figure 7.8).

Move icon

24. Pick the mid-point of the selected line. Move the mouse towards the left – while locked to the horizontal (Figure 7.8). **Do not click yet.**

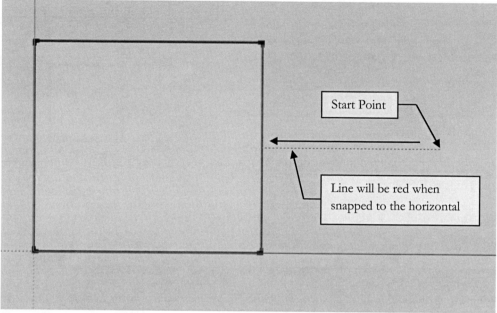

Start Point

Line will be red when snapped to the horizontal

FIGURE 7.8 Moving an edge

25. While in the *Move* command and snapped to the horizontal (i.e., red axis), type **1'6** and then press **Enter**.

That is it! You essentially just stretched the rectangle. The two horizontal lines automatically shrunk in length and the surface resized itself as well.

26. **Save** your *Night Table.skp* file.

file name: **Small Dresser**   *file name:* **File Cabinet**

27. Draw the *Small Dresser* and the *File Cabinet* per the previous instructions.

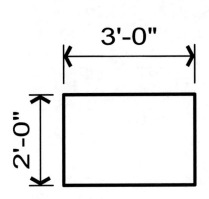

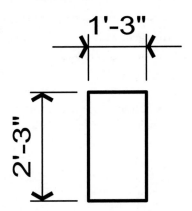

file name: **Square Chair**

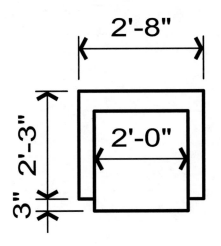

Next you will draw this squarish-styled chair. You could draw this by setting up a few *Guides*, but another method will be shown. It is good to know several ways to accomplish the same thing as one solution may be more efficient than another in certain situations.

28. Start with a 2′x2′ square aligned with the origin.

29. Draw the backrest and armrests as separate rectangles, near the square (Figure 7.9)

30. Use the **Move** tool to move the rectangles into position (Figure 7.10).

SketchUp does not allow overlapping lines. Therefore, when you moved the rectangles into place the lines were merged, with any endpoints remaining. This will allow you to delete the extra lines identified in Figure 7.10.

31. **Delete** the extra lines identified in Figure 7.10; select it and press the **Delete** key on the keyboard.

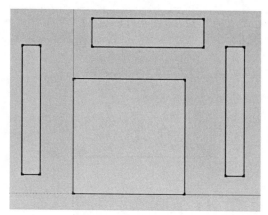

FIGURE 7.9 Creating the chair

When the two lines were removed, the surfaces automatically joined together within the newly enclosed area (Figure 7.11). At this point you only have two surfaces defined, each of which can be extruded into separate shapes, which you will do later in the next chapter.

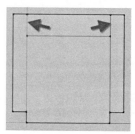

FIGURE 7.10

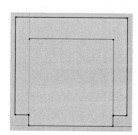

FIGURE 7.11

You have completed the square chair – for now anyway. You can save your file and move on to the next one.

Selecting Objects

At this time we will digress and take a quick look at the various techniques for selecting entities in SketchUp. Most tools work the same when it comes to selecting elements.

When selecting entities, you have two primary ways to select them:

o   Individually select entities one at a time
o   Select several entities at a time with a window

You can use one or a combination of both methods to select elements when using the *Select* tool.

Individual Selections:

When using the *Select* tool, for example, you simply move the cursor over the element and click; holding the **Ctrl** key you can select multiple objects. Then you typically click the tool you wish to use on the selected items. Press **Shift** and click on an item to subtract it from the current selection set.

Window Selections:

Similarly, you can pick a *window* around several elements to select them all at once. To select a *window*, rather than selecting an individual element as previously described, you select one corner of the *window* you wish to define. That is, you pick a point in "space" and hold the mouse button down. Now, as you move the mouse you will see a rectangle on the screen that represents the windowed area you are selecting; when the *window* encompasses the elements you wish to select, release the mouse.

You actually have two types of windows you can use to select. One is called a **window** and the other is called a **crossing window**.

Window

This option allows you to select only the objects that are completely within the *window*. Any lines that extend out of the *window* are not selected.

Crossing Window

This option allows you to select all the entities that are completely within the *window* and any that extend outside the *window*.

Using Window versus Crossing Window

To select a *window* you simply pick and drag from *left to right* to form a rectangle.

Conversely, to select a *crossing window*, you pick and drag from *right to left* to define the two diagonal points of the window.

file name: Square Sofa

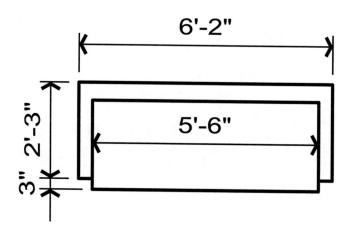

Similar to how you turned the *Small Desk* into a *Night Table*, you will turn the *Square Chair* into this *Square Sofa*.

This is a little trickier as you have to select multiple lines as part of the *Move* command.

32. Once you are sure you saved the *Square Chair* file, do a **Save As** to create the **Square Sofa** file.

33. Select the three lines shown in Figure 7.12.

Moving these lines will cause the horizontal lines to follow – or stretch, thus growing the chair into a sofa!

34. Select **Move** from the toolbar.

35. Pick anywhere in the drawing area and begin moving the cursor to the right, snapped to the horizontal (i.e., on red axis).

36. Without clicking a second point, type the desired length you wish to move the lines (i.e., stretch the sofa); this is the difference between the chair and the sofa.

Notice how the horizontal lines extended because they have a parametric relationship to them. The surface has also updated.

FIGURE 7.12

37. Use the *Tape Measure* tool to double-check your dimensions. Make any corrections needed before moving on.

Keep in mind that all these basic steps will be directly applicable to the more advanced 3D modeling coming up.

file name: **Range**

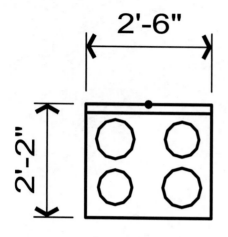

2'-6"

2'-2"

Now you will draw a kitchen range with four circles that represent the burners.

In this exercise you will have to draw temporary lines, called *Guides*, to create reference points needed to accurately locate the circles. Once the circles have been drawn the *Guides* can be erased.

38. In a new file, **Draw** the range with a 2″ deep control panel at the back; refer to the steps previously covered if necessary.

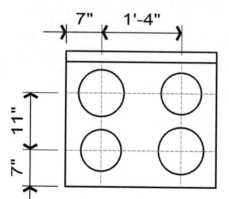

7" 1'-4"

11"

7"

FIGURE 7.13 Adding guides

39. Use the ***Tape Measure*** tool to create the four guides dimensioned in Figure 7.13.

    a. With the *Tape Measure* tool active, click and drag on an edge to create a *Guide* parallel to that edge.

    b. Once you let go of the mouse the *Guide* is created.

    c. Once placed, type the distance and then press **Enter** to position it.

40. Using the ***Circle*** tool, draw two 9½″ Dia. circles and two 7½″ Dia. circles using the intersections of the *Guides* to locate the centers of the circles (Figure 7.13).

Circle icon

41. Using the **Select** tool, select each of the *Guides* and press the **Delete** key on the keyboard to remove them from the model.

Circles are made up of several straight line segments. The default number is 24. If your circle is large, the number of edges needs to be increased to maintain the look of a circle. This can be done just before clicking to locate the circle's center, by just typing a number and pressing **Enter**. This number, the radius and *Layer* can all be changed at any time using the *Entity Info* dialog.

file name: **Rounded Chair**

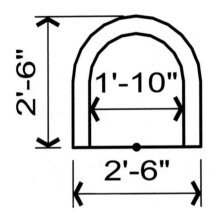

These last two drawings will involve using the *Arc* tool. The *Offset* command will also be utilized.

You will start this chair drawing by sketching the perimeter and then offsetting it inward.

42. Draw the three orthogonal lines shown in Figure 7.14.

    a.   The horizontal line is centered on the green axis and aligned with the red axis.

    b.   The two vertical lines are 1'-3" long.

43. Use the **Arc** tool to add the rounded backrest.

Arc icon

    a.   Pick the three points shown in Figure 7.14.

    b.   When picking the third point you should see a "half circle" tooltip before clicking the mouse button.

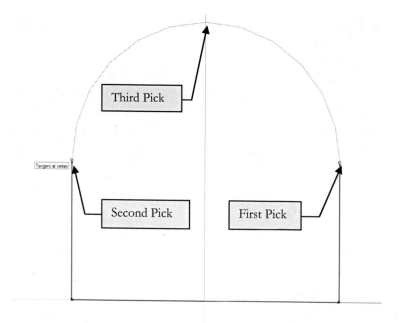

Second Pick — Third Pick — First Pick — Tangent at vertex

FIGURE 7.14 Sketching an arc

You now have the perimeter of the chair defined. Next you will use the *Offset* command to create the backrest.

44. Select the **Offset** tool from the toolbar.

45. Select the arc – just the line, not the surface.

Offset icon

46. Start moving your cursor towards the middle of the chair and then type **4** and press **Enter**.

47. Draw the two remaining vertical lines.

  a. For the first end of the line, snap to the end of the arc.

  b. For the second end of the line, snap to a perpendicular point on the horizontal line.

The 2D version of the chair is now complete. The drawing now has two surfaces, because two enclosed areas exist; the backrest and the seat area. This will make the 3D extrusion process go smoothly.

file name: **Love Seat**

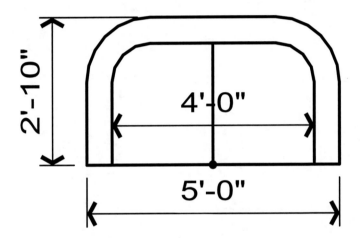

This last 2D exercise will round things off by making quarter round arcs.

One thing that is a little tricky here is setting up points to pick before sketching the arc. When adding an arc, it is easiest if the two endpoints are defined by other line work. Thus, you just have to snap to endpoints for your first two picks and the third defines the radius. In this example you will need to add two temporary lines to define the start and end points of the arc.

The radius of the two arcs is 1'-3".

48. Add the orthogonal lines shown in Figure 7.15.

    a.   All dimensions can be deduced from the information given.

The two 1'-3" lines are temporary and have only been added to aid in sketching the arcs (which is done in an upcoming step).

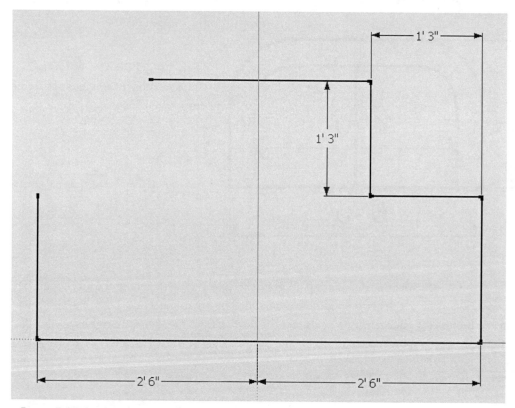

FIGURE 7.15 Setting things up for the arc tool

49. **Delete** the two 1'-3" lines.

50. Add one of the perimeter arcs.

    a. Pick the two endpoints provided by the orthogonal lines.

    b. Move the cursor until the arc is purple and the tooltip reads "Tangent at vertex" – this will produce a quarter round arc.

51. Add the other arc per the previous step.

Notice that you did not have to specify the radius of the arc because of the preparation done.

52. Offset the two arcs inward **4"** – similar to the rounded chair (Figure 7.16).

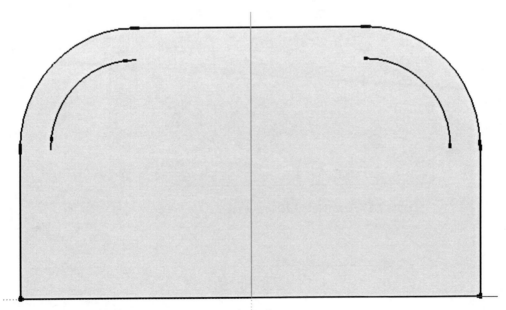

FIGURE 7.16 Offset perimeter arcs inward 4 inches

53. Complete the backrest and armrests by "connecting the dots" via the *Line* tool.

54. Draw the vertical line down the middle, representing the cushions.

55. **Save.**

Sometimes it is better to extrude a shape into the third dimension before adding extra line work such as the cushion line. This issue will be explored more in a later lesson.

Your drawing should be complete (Figure 7.17). Note that the drawing has three surfaces based on the line work you sketched. Don't forget to save your file.

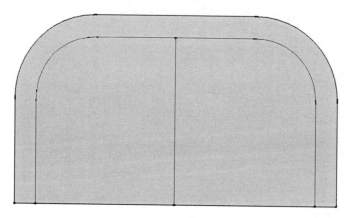

FIGURE 7.17 Completed 2D love seat

Chapter 7 Review Questions

The following questions may be assigned by your instructor as a way to assess your knowledge of this section. Your instructor has the answers to the review questions.

1. **46.5** is the same as **3' 10 1/2** when entering values in SketchUp. (T/F)

2. When a line/edge is moved, the adjacent, intersection, lines not will grow or shrink to stay attached. (T/F)

3. If the "foot" or "inch" symbol is not entered inches are assumed. (T/F)

4. You can turn on additional toolbars via **View → Toolbar...** (T/F)

5. Sometimes it is better to extrude a shape into the third dimension before adding extra line work. (T/F)

6. When adding an arc, it is easiest if the two endpoints are defined by other line work. (T/F)

7. Selecting from right to left is a _____ Window.

8. Tape Measure results are listed in the _____ _____ _____ .

Chapter 8
3D Modeling

In this chapter you will transform the objects created in the previous chapter into three dimensional models. Once the 3D objects have been developed, you will learn how to apply materials, in the next chapter, to make them look more realistic.

file name: **Bookcase – 3D**

You will open the files from the previous chapter and *Save As* to a new file name so the original files remain intact (just in case you need to start over).

1. Open the **Bookcase.skp** file created in the previous chapter.

2. Select **File → Save As**.

3. Save the file as **Bookcase – 3D**. (You may save it into another folder if you wish.)

Next you will turn on the *Views* toolbar to make it easier to switch between 3D and elevation/plan views.

4. Select **View → Toolbars → Views**.

5. Click on the **3D** icon on the *Views* toolbar.

You should now be seeing a 3D view of your bookcase drawing (Figure 5-5.1). Of course it is still just a 2D drawing being viewed from a 3D vantage point. Also recall that the view mode of the model is currently set to *Parallel Projection* rather than the default *Perspective* mode. You will leave it this way for now.

Using the Push/Pull Tool

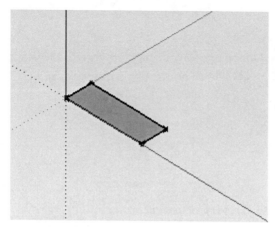

Now you will use the **Push/** l to quickly turn the 2D outline into a 3I ased) object. This tool requires you to select a surface, which means an enclosed perimeter is required. The tool takes the previously defined 2D shape and extrudes it into a 3D shape. So the end result is several additional edges and surfaces which give the appearance of a 3D object.

6. From the *Getting Started* toolbar, select the **Push/Pull** tool.

FIGURE 8.1 Parallel Projection 3D view

Next you will simply click and drag on the surface portion of your bookcase (i.e., rectangle) and start moving the mouse straight upwards. You will not worry about an exact height as the precise value will be typed in immediately after using the *Push/Pull* tool.

7. Move your cursor over the surface until it pre-highlights and then **click and drag the mouse up**.

You can only drag in a direction perpendicular to the surface – in this case, straight up or down.

8. **Release the mouse** button at any time, once the 3D geometry appears in the correct direction (i.e., up versus down below the ground).

9. Immediately after releasing the mouse button, type **4′** and then press **Enter**.

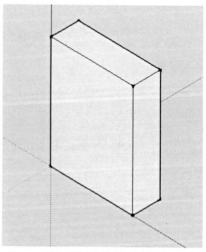

FIGURE 8.2 4′ high extrusion

You now have a 4'-0" high extrusion which will be further refined to look like a bookcase (Figure 8.2).

In addition to the *Push/Pull* tool making solid (or additive) geometry, it can also be used to make voids (or subtractive) geometry. That is, if you sketch rectangles on the face of the bookcase and then use *Push/Pull* you can create a void by "pushing" into the larger 3D object. Conversely, if you "pull" the rectangular shape you would create a bump-out on the face of the bookshelf.

The following steps can be done in the current 3D view, and are easier once you get used to working in a 3D view. You can try following the next few steps in the current 3D view or you can switch to the front view (per the very next step) to follow along exactly with the book.

10. Click the **Front** icon on the *Views* toolbar.

You are now looking at the equivalent of a 2D elevation. This is only true if you are still in *Parallel Projection* mode; if not, you see more of a 3D view of the front.

11. Select the **Offset** tool from the toolbar.

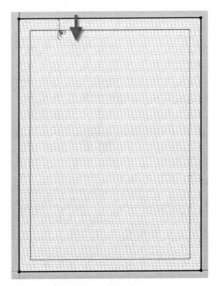

Before you click and drag on the edge you want to offset, you need to see the surface highlight first. This lets you know the edge will offset on the correct plane/direction.

12. Move your cursor over the surface. Once you see it highlight, click **near** the top edge, drag your cursor downward, and let go of the mouse button (Figure 8.3).

FIGURE 8.3 Offset on front face

13. Type **1″** and then press **Enter** to make the newly created line work one inch away from the perimeter.

14. Click the **Select** tool, and then select the bottom edge of the newly created rectangle.

15. Use the *Move* tool and reposition the line **3″** up (Figure 8.4).

   a. In the *Front* (or *Iso*) view, click the **Move** tool

   b. Click to select the line

   c. Move your cursor upward, snapped to the blue axis

   d. Click again to move the line (don't worry about the exact location)

   e. Type **3″** and press **Enter**

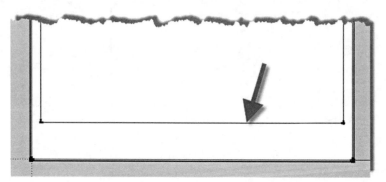

FIGURE 8.4 Line moved, 4″ from bottom edge

Notice the pattern when it comes to specifying lengths; first get the entity generally in the right spot and then enter a specific length. You need to enter the value immediately after the geometry has been created or edited.

Now you will sketch the line work for the shelves.

16. Sketch the two shelves (Figure 8.5):

    a. Each shelf is **1″** thick.

    b. The shelves are **equally spaced**.

    c. *To make the shelves equally spaced:* Use the **Tape Measure** tool to list the overall distance within the inner rectangle. <u>Do the math:</u> overall distance minus 2" (for the shelves) divided by three.

    d. Copy the top or bottom line using **Move** (while holding down the **Ctrl** key to get into *Copy* mode).

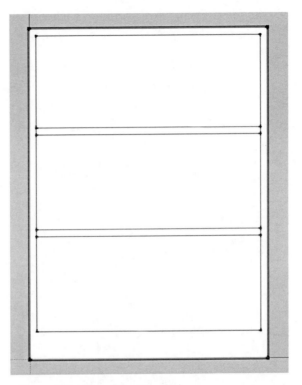

FIGURE 8.5 Lines added for shelves

As you copied the horizontal line up, a new perimeter was defined on the front surface. So SketchUp created a new surface, and modified the previous surface.

Now it is time to switch back to the 3D view and use the *Push/Pull* tool to carve out the shelf areas. The large rectangular surfaces you are about to push will become the inside face of the back panel for the bookshelf. If that did not make sense, it will in a moment.

17. Switch back to the 3D view.

18. Select **Push/Pull** and then click and drag on the larger top rectangle (Figure 8.6).

19. Drag the mouse about halfway into the bookcase and let go of the mouse.

20. Type **11″** and then press **Enter**.

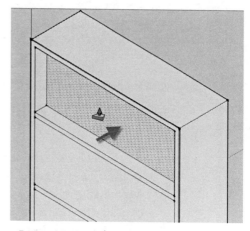

FIGURE 8.6 Push/Pull void

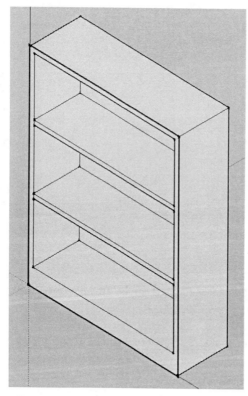

FIGURE 8.7 Final bookcase

The book case is 12″ deep, so pulling the surface back 11″ gave us a 1″ thick back panel.

21. Repeat this process for the other two larger rectangles (Figure 8.7).

The final bookcase should look like Figure 8.7. You can press and hold down your center wheel button and orbit around the bookcase to see it from various angles. When finished, just click the **3D** icon again to reset the view.

22. **Save.**

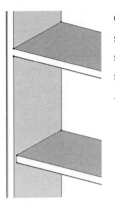

One last comment on the bookcase (Figure 8.8): If you want the face of the shelf to appear flush with the edge panel you can zoom in and erase the small vertical line (top shelf example). Or, if you want the shelf to appear separate or recessed slightly, you can leave the line in place and use the *Push/Pull* tool (bottom shelf example).

FIGURE 8.8 Shelf options

file name: Coffee Table – 3D

Using similar techniques to developing the bookcase, you will now create a coffee table.

23. Open the 2D coffee table model and do a **Save As** to **Coffee Table – 3D**.

24. Use the ***Push/Pull*** tool to make the 2D rectangle **2′-0″** high.

    a.   See the previous steps for more information on how to do this.

25. Switch to the **Front** view.

This will be a heavy mass looking coffee table. The next steps start to define the thickness of the top and legs as viewed from the front.

26. Using the ***Line*** tool, sketch the 5 lines shown in Figure 8.9.

    a.   Create 3″ wide legs and a 3″ thick top.

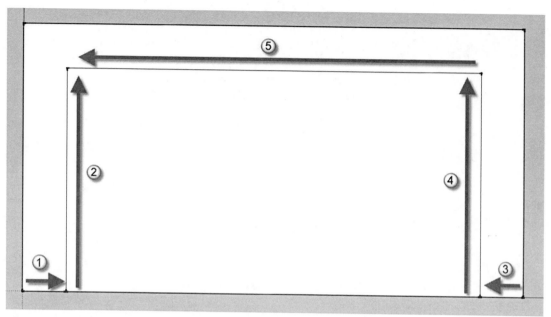

FIGURE 8.9 Coffee table front view with lines sketched

SketchUp does not allow two lines to overlap. So when you sketched lines 1 and 3 the original line across the bottom split into smaller line segments based on where your new line stopped. This is good as it keeps the file less cluttered and it ultimately creates a large surface in the middle; we will use the *Push/Pull* tool to carve out that area.

Keep in mind this could have been done from a 3D view. Once you are more experienced, it will be more efficient to do this in a 3D view. The main trick is making sure you stay on the correct *Axes*.

Next you will define the two side views in the same way the front view was developed. You have to do this for both sides, but not the back view. You will see why in a moment.

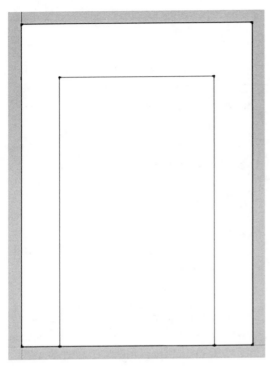

FIGURE 8.10 Right view with lines added

27. Switch to the **Right** view and add the 3″ wide legs and a 3″ thick top, per the previous steps (Figure 8.10).

28. Repeat the previous step for the **Left** side.

You are now ready to switch back to a 3D view and use the *Push/Pull* tool to remove the large mass below the table.

29. Switch to the **3D view**.

30. Use the ***Push/Pull*** tool to select the large rectangular area on the front of the coffee table.

31. Push the surface back into the model until it "snaps" into alignment with the back face of the model. Release the mouse button.

The model should look like Figure 8.11. If not, you can either *Undo* and try again, or grab the surface again using the *Push/Pull* tool. You may need to *Orbit* a little to get a better view (but this should not be necessary). Notice, when the two faces (front and back) come together, SketchUp deleted both and the end result is a void.

If you would have done one of the sides first, you would get the same result. However, now the sides will be a little different due to the top edge now being defined on the back side of the legs. In this case, you will end up needing to manually delete one surface and a line.

32. Use **Push/Pull** on one of the side of the table, pushing the surface in to align with the back side of the legs (Figure 8.11).

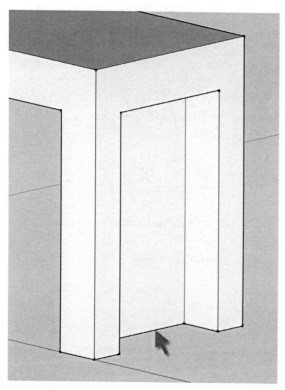

Notice the surface and bottom edge did not automatically get deleted when modifying the side view (Figure 8.11). Again, this is due to the surface ending directly on the edge of the underside of the 3″ thick top (the top edge was created when the front was extruded back).

This is an easy fix.

33. Use the **Select** tool to pick the bottom edge (i.e. line) and press the **Delete** key.

Once the perimeter is gone the surface must go as well. You could have deleted the surface and then the line but that would be more work!

FIGURE 8.11 Right side adjusted

That concludes the coffee table exercise. Be sure to save as a new file name per the first step for this model. Use *Orbit* and *Tape Measure* to inspect your work!

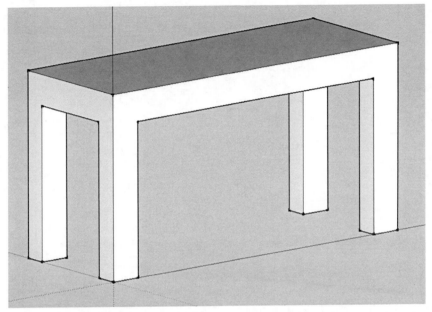

FIGURE 8.12 Completed coffee table

file name: **Small Desk – 3D**

Here you will continue to build on the concepts covered thus far.

34. **Open** the *Small Desk* file and **Save As** to **Small Desk – 3D**.

35. Extrude the rectangle to be **30″ high** using *Push/Pull*.

36. Switch to the front elevation view and add the line work shown in Figure 8.13. Use the ***Line*** tool and **Move + Ctrl** to *Copy* lines rather than *Offset* to try a different technique. Be sure to erase any extra lines before moving on.

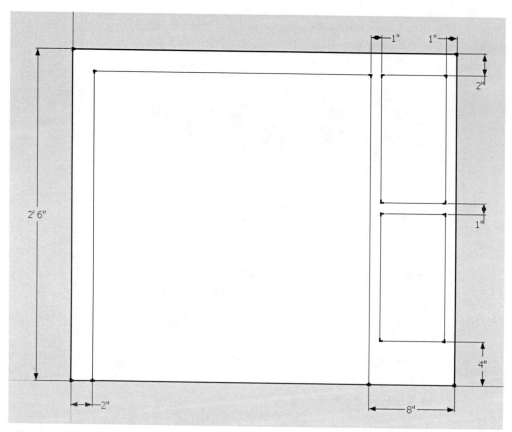

FIGURE 8.13 Front elevation of small desk

Next you will carve out the leg space and bump out the drawer panels. Notice you will not sketch in the pulls (i.e., handles) for the drawers yet. If the handles (i.e. pulls) were sketched now you would have another face that would not *Push* or *Pull* with the rest of the drawer panel. Instead, you will get the drawer panel extruded and then sketch the pull outline on the face of the new panel surface.

37. Use **Push/Pull** to modify the desk (Figure 8.14):

  a. Knee space should be 1'-4" deep.

  b. Drawer panels should be ½" thick.

Your model should look like this (Figure 8.14).

Next, you will create the pull for the top drawer. This will be a simple pull which will be turned into a *Component* and copied down to the lower drawer.

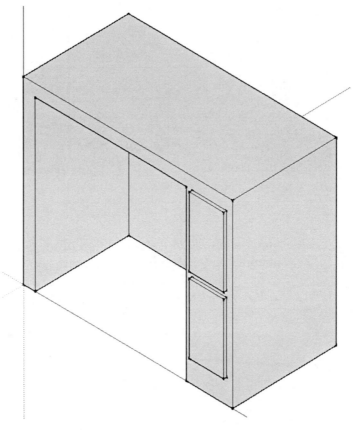

FIGURE 8.14 Modified desk

38. Create the rectangle for the pull on the face of the drawer panel. Exact size and location are not critical; just try and get it close (Figure 8.15).

39. Use **Push/Pull** to extrude the pull out from the drawer panel (Figure 8.16).

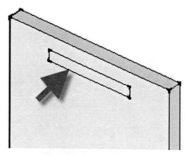

FIGURE 8.15 Pull outline

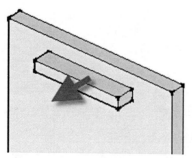

FIGURE 8.16 Pull Extruded

40. Sketch the profile of the pull on top of the extruded shape. This is in the 3D view; wait until you see the tooltip which says "on face" for drawing (Figure 8.17).

41. Use **Push/Pull** to carve out the back side of the pull (Figure 8.17).

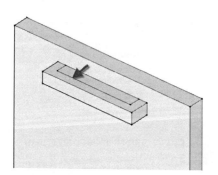

FIGURE 8.17 Pull profile

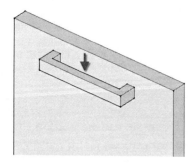

FIGURE 8.18 Pull completed

Next, you will group all the lines and surfaces that make up the pull into a single entity called a *Component*. As mentioned previously in this book, using *Components* is more efficient and makes the file smaller when the item will be used many times.

42. In the 3D view, *Zoom* in on the pull and select a window picking from left to right.

   a. Be sure to **pick from left to right** and adjust the view so you are only selecting the pull and nothing else behind it.

43. With the pull selected, click the **Make Component** tool

   FYI: Turn on the Principal toolbar or right-click.

44. Fill out the dialog as shown in Figure 8.19.

45. Click **Create**.

You now have a *Component* created in the model of your pull. Because the "Replace selection with component" was selected, the original lines and surfaces have been replaced with a copy of the new *Component*.

Next you will copy the *Component* down to the other drawer.

46. Select the *Component* and use the **Move + Ctrl** key to copy the pull down to the other drawer. Use the pick points shown in Figure 8.20.

FIGURE 8.19 Create Component dialog

Now that the pull is a *Component* you can edit one and the other will instantly update. You will try editing a *Component* next.

47. **Right-click** on one of the pulls and then select **Edit Component** from the pop-up menu.

48. Make a change to the pull, something simple such as using ***Push/Pull*** to make the extrusion taller (i.e., thicker) or something more detailed like making the pull curved (Figure 8.21).

49. When finished editing the *Component*, pick the *Select* tool, and then click away from it and the edit mode is finished.

Notice both pulls have been updated. This can save a lot of time!

Another right-click option is *Explode*. This allows you to reduce a *Component* back down to its basic elements and be changed differently from the other *Components*. *Explode* only affects the selected element(s).

50. **Save** your model.

Before leaving the small desk, you will take a look at one more thing. As SketchUp models get more complicated, they become harder to modify. However, the process follows the same general steps as with the 2D shapes.

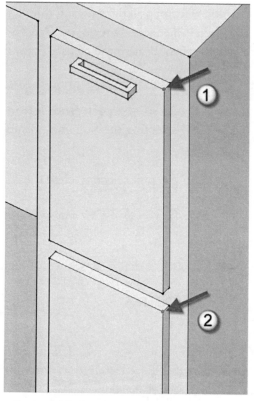

FIGURE 8.20 Copying component

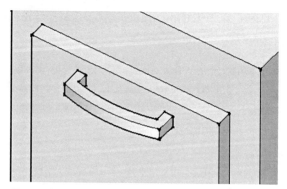

FIGURE 8.21 Editing a component

Next, you will make the desk 8″ wider because the drawers are too narrow.

51. Switch to the front view.

52. Drag a selection window, going from left to right (Figure 8.22).

    a. This will not work properly if you pick in the opposite direction (i.e. from right to left).

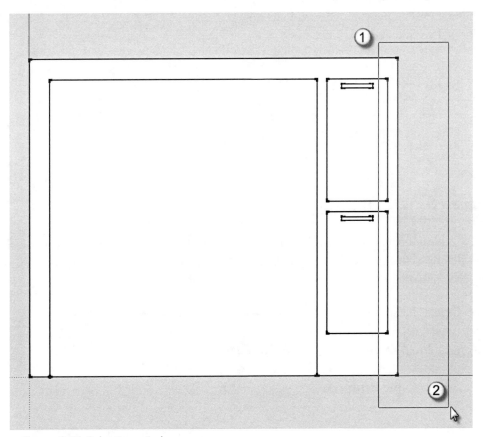

FIGURE 8.22 Selection window

53. **Move** the selected entities **8″** to the right.

54. Select the two pulls and move those **4″** to the right.

55. Switch to the **3D view**.

The model is now modified (Figure 8.23)! Notice how the faces and related lines stretched to following to repositioned vertical lines.

56. **Save**.

FIGURE 8.23 Complete

file name: **Small Dresser – 3D**

Often, it is ideal to add *Guides* to define an area rather than starting with lines. This is because lines can divide other lines and surfaces. You will use *Guides* to define the location of the drawer panels before sketching them.

57. **Open** the small dresser file and **Save As** to **Small Dresser – 3D**.

58. Use *Push/Pull* to make the dresser **4′-6″** tall.

59. Switch to the **Front** view.

60. Add the ten **Guides** shown in Figure 8.24.

    a. To add a *Guide*: click and drag on an edge using the *Tape Measure* tool; and then type a distance for accuracy.

    b. The bottom *Guide* is **4″** up from the floor.

    c. All remaining spacing is **1″** and the drawer panels are equal in height.

    d. Do the math first, to determine what the panel height should be.

FIGURE 8.24 Guides added

It is worth pointing out that SketchUp has a **Divide** tool, but it will not work in this case. To use it, you draw a line and then right-click on it, select **Divide** and then type in a value and press **Enter**, but this does not take into account the spacing between the drawer panels

With the guides in place, you can quickly sketch the rectangles to define the edges of the drawer panels.

61. Use the **Rectangle** tool to define the perimeter of the drawer panels; simply snap to the intersections of the *Guides*.

62. Switch to the **3D view.**

Your model should look like Figure 8.25. Note that the *Guides* were created on the plane related to the edge used to define it. Thus, picking the grid intersections causes the rectangles to be created on the correct plane – which subsequently divides the surface. You are now ready to *Pull* the panels.

63. Use **Push/Pull** to extrude the panels ½″ out from the surface.

You could leave the *Guides*, and even turn them off so they are not in the way. However, you will just delete them as they are no longer needed.

64. Select the **Eraser** tool and erase each of the *Guides* (by picking them).

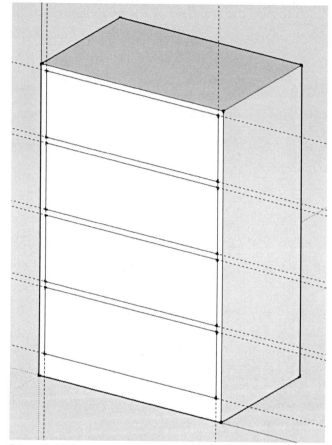

FIGURE 8.25 Rectangles added

The last step in completing the dresser is to add the pulls. You could recreate them from scratch, following the previous steps covered. However, it would be much faster to open the Small Desk – 3D file and copy one of the pulls to the *Clipboard*, and then *Paste* it into the dresser model. You will see that the pull will want to automatically "snap" to a surface due to the "glue to" setting when the *Component* was created.

65. **Open** the **Small Desk – 3D** file

    a. If you browse to the file, using *Windows Explorer*, and then double-click the file, it will open another session of SketchUp. This will allow you to keep the dresser file open.

66. Select one of the pulls and press **Ctrl + C** to *Copy* it to the *Clipboard*.

67. Switch back to the dresser model and press **Ctrl + V** to *Paste* the pull into the current model.

Notice as you move your cursor around the screen, the pull follows the surface below your cursor. It wants to stick to a surface due to the "glue to" option setting when it was first created.

68. Place the pulls approximately as shown in Figure 8.26.

You may want to switch to a "front" view and then add *Guides* to get the first drawer set up. Once you have one drawer set, you can copy it from one drawer to the next until they all have pulls.

69. **Save** your model.

The dresser is done for now.

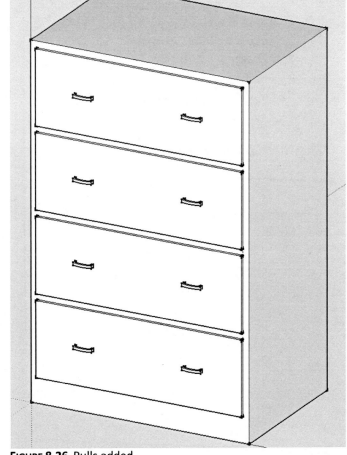

FIGURE 8.26 Pulls added

file name: **File Cabinet – 3D**

It is not always necessary to use the *Push/Pull* tool to create an extruded 3D element to represent something. In the case of this file cabinet it is easier to simply sketch the lines on the face of the cabinet which nicely defines the two drawers. This model can be developed more quickly and is less of a burden on the overall model due the less complex geometry and fewer faces and surfaces.

Worrying about the complexity of this file cabinet may seem trivial now, but in a large model with several chairs, desks, file cabinets, etc., you can see how the overall number of faces and edges could really start to bog down even the fastest computer.

70. **Open** the *File Cabinet* file and **Save** it as **File Cabinet – 3D.skp**.

71. Develop the cabinet:

   a. **27″** high

   b. Pulls **Copy/Pasted** from previous file.

   c. Use the **Line** tool to sketch the drawers:
      i. 1″ space on sides and top;
      ii. 4″ space from bottom;
      iii. Drawers are equal height.

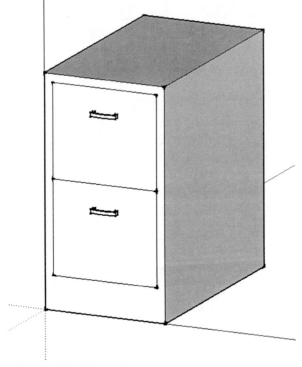

FIGURE 8.27 File cabinet

Your file cabinet should look like Figure 8.27. Even though new surfaces were made within the perimeter of the drawers, they do not need to be extruded into a 3D element.

72. **Save.**

At this point you have learned to use a majority of the tools typically used on a regular basis by the average SketchUp user. There certainly is more to learn, but you are well on your way!

file name: **Square Chair – 3D**

In this exercise you will learn to modify geometry once it is 3D; making the backrest higher than the armrests.

73. **Open** the *Square Chair* file and **Save** it as **Square Chair – 3D.skp**.

74. Use ***Push/Pull*** to make the armrest and backrest **26″** high (Figure 8.28).

75. Make the seat area **16″** high (Figure 8.29).

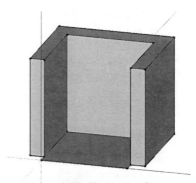

FIGURE 8.28 First extrusion

You now have the basic chair defined. However, you decide you would like to have a higher backrest and have the armrests curve down towards the front.

76. *Orbit, Pan* and *Zoom* in as needed, and then use the ***Line*** tool to add the two lines shown in Fig. 8.30.

77. Use ***Push/Pull*** to make the back rest **4″** higher.

78. Select and erase the extraneous lines highlighted in Figure 8.31.

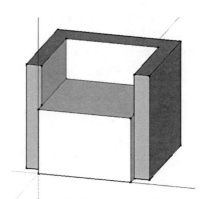

FIGURE 8.29 Second extrusion

Notice how the surface automatically heals itself when the lines are deleted.

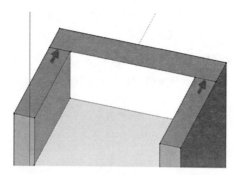

FIGURE 8.30 Lines added at backrest

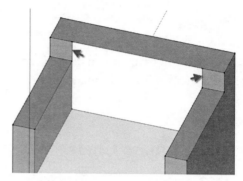

FIGURE 8.31 Lines to be erased

Now you will carve out a portion of the arm rests to make them curve down and toward the front of the chair. You will switch to a side view (this can also be done in a 3D view). In the side view you will sketch the profile of the portion you wish to exclude. The *Push/Pull* tool will be used to carve out the portion not needed, just like you did with the coffee table. But, before you use the *Push/Pull* tool you will want to copy the profile over to the other armrest so you do not have to recreate it.

79. Switch to a side view and sketch an **Arc** similar to that shown in Figure 8.32.

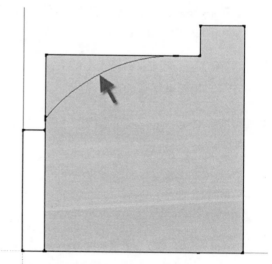

FIGURE 8.32 Adding arc to armrest

80. Switch back to the **3D view** and *Copy* the arc to the other armrest (Figure 8.33).

81. Use **Push/Pull** to drag the new surface over to align with the other side of the armrest – this will cause the entire extrusion to be deleted; repeat for other side.

As you can imagine, these types of modifications could continue to be made on this chair until it was exactly the way you want it. You may want to do a **Save As** once in a while to make it easy to go back to a previous design state if you get to a point where you are not happy with the design and want to go back.

Keep in mind that edges are required anytime there is a change in direction of adjacent surfaces, but not when surfaces are coplanar. So, the line at the top of the curved armrest is required in Figure 8.34, but the lines you deleted in Figure 8.31 were not.

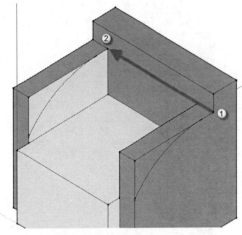

FIGURE 8.33 Copy arc to other armrest

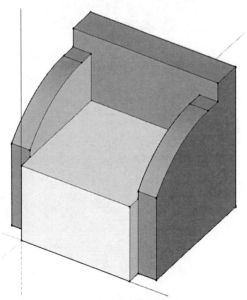

FIGURE 8.34 Use Push/Pull on armrest

Project Example: Custom Furniture.

Image courtesy of LHB; www.LHBcorp.com

Chapter 8 Review Questions

The following questions may be assigned by your instructor as a way to assess your knowledge of this chapter. Your instructor has the answers to the review questions.

1. SketchUp does not allow two lines to overlap. (T/F)

2. You need to use the Move tool when you want to make a copy. (T/F)

3. SketchUp does not have an icon for creating a component. (T/F)

4. SketchUp can divide a line into a specified number of segments. (T/F)

5. For a surface to exist, edges are required anytime there is a change in direction of adjacent surfaces. (T/F)

6. Which tool quick extrudes a shape into a 3D object: _____ .

7. Command that saves a copy of your current model within SketchUp: _____.

8. Sometimes it is better to sketch 2D lines on a surface rather than extruding everything to 3D. (T/F)

Chapter 9
Working with Materials

Now that you know how to develop simple geometry, you will make the models look a little more realistic and add materials to them. SketchUp comes with a nice selection of materials from which to choose. It is also possible to scan a material and use it on your model!

Materials Dialog

The *Materials* dialog is used to select and add *Materials* to your model.

1. **Open** your SketchUp model: **Bookcase – 3D.skp**.

2. Select the **Paint Bucket** icon from the toolbar.

You are now in the *Paint Bucket* tool, and if it was not already open, the ***Materials*** dialog box pops up.

3. Pick **Wood** from the drop-down list in the *Materials* dialog box (Figure 9.1b).

At this point you have ten wood materials from which to choose. Most of these are flooring materials (thus, they have lines in them to represent floor boards) and one material is OSB (Orientated Strand Board) which is not usually a finish material.

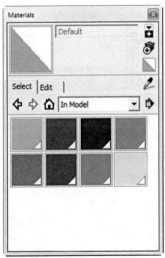

FIGURE 9.1A Materials

FIGURE 9.1B Materials list

4. Select **Wood_Cherry_Original** from the options listed under *Wood*.

5. Move your cursor into the model area and click on 2 different surfaces to apply the material (only click on two for now). Figure 9.2

As you can see, the material is added to each surface as you click on it using the *Paint Bucket* tool. This is handy if you want to have the shelves be a different material than the frame of the unit. Similarly, if you need to add more refined materials you can sketch additional lines to divide the surfaces into separate areas.

However, clicking each surface would take a bit of time. So the next steps will show you how to quickly add the material to the entire bookcase at once.

6. With the *Paint Bucket* tool still active, right-click on one of the surfaces and pick **Select → All Connected**; the same *Layer* would also work in this case (Figure 9.2).

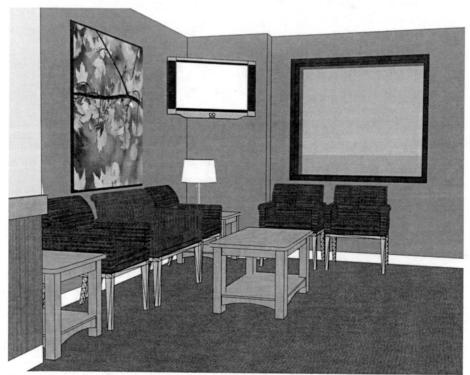

Project Example: Waiting Room.   *Image courtesy of LHB; www.LHBcorp.com*

7. Click on one of the surfaces to apply the material to all faces.

> ***TIP:*** *You can also press* CTRL *or* SHIFT, *while painting a surface, to…*
> ***CTRL:*** *Paint all connected faces with matching paint*
> ***SHIFT:*** *Paint all faces in model with matching paint*

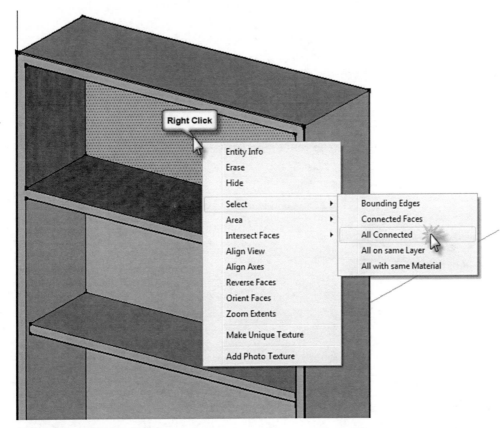

FIGURE 9.2 Adding materials

The wood material has now been added to the entire bookcase. The materials are designed to be real-world scale, so you need to model things the same size they would be in the real world. Otherwise, things such as bricks will not look right.

8. **Save** your **Bookcase – 3D** file before moving on.

In the next steps you will add a transparent material to give the appearance of glass. This will be a glass panel in the center of the coffee table.

9. **Open** the *Coffee Table – 3D* file.

The first thing you will do is carve out an area for the glass panel in the center of the table.

10. Switch to the **Top** view and *Offset* the outer edge inward **3″**.

11. Switch back to the **3D view** and use *Push/Pull* to remove the center of the table; click and drag the surface down until it aligns with the bottom edge of the table top.

12. Use the *Select* tool to pick the remaining surface and **Delete** it.

Your model should now look like Figure 9.3.

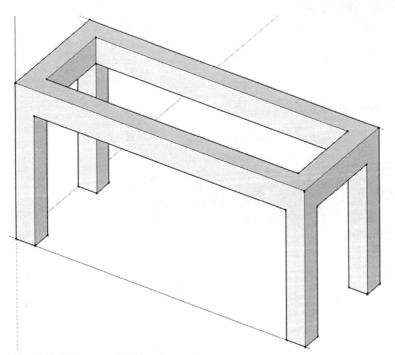

FIGURE 9.3 Center of table removed

Before modeling the glass, now is a good time to quickly make everything a wood material.

13. Using the steps recently covered, make all elements material: **Wood_Board_Cork**.

The material may not actually be cork, but the color might be the closest thing to what you are thinking, and that will work for now.

Next, you will create a surface at the top of the new opening to represent the glass. This can easily be done in the **3D view**.

14. Select the **Rectangle** tool and then pick two opposite corners of the opening in the center of the table (Figure 9.4).

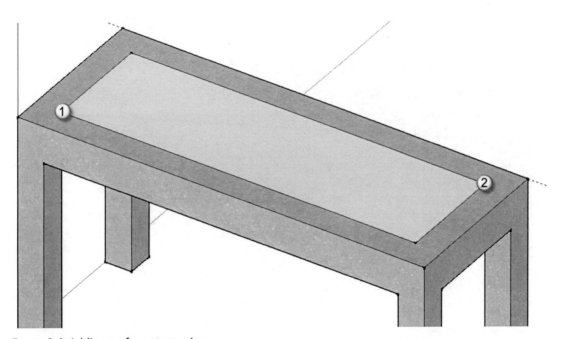

FIGURE 9.4 Adding surface at opening

As mentioned earlier, SketchUp does not allow lines to overlap. So the lines for the rectangle were immediately deleted, but they caused SketchUp to check and see if any enclosed areas need a surface.

15. Use the *Paint Bucket* tool to set the new surface to **Translucent \ Translucent_Glass_Corrugated**.

The glass material is now added to the coffee table (Figure 9.5). There are a number of things you could do at this point. You could select the surface and copy (via **Move + Ctrl**) the surface down ½″ inch to give the appearance of thickness.

You can also right-click on the glass surface, and select **Texture → Position**. This gives you the option to rotate and adjust the scale of the material.

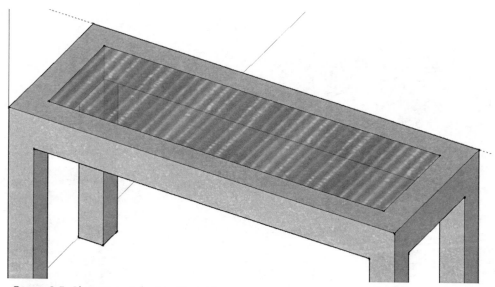

FIGURE 9.5 Glass material added to table

Just in case you changed something, the following needs to be set so your screen matches the screen shots in this book.

- View menu
  - Edge Style
    - Edges (checked)
    - Profiles (checked)
    - Extensions (checked)
  - Face Style
    - Shaded with Textures (checked)

Try **X-ray** and **Back Edges** before moving on, and then set things back to the defaults listed above.

As you did with the desk, try adjusting the overall length and width of the table. You have to use very specific selection windows to make this type of modification.

16. **Save** your model.

This concludes the basic introduction to SketchUp. You will learn a few new things later in the book, when you start the design of the interior fit out. However, you are already off to a good start.

17. Use the techniques learned here to create the remaining 3D models; based on the 2D sketches created in the previous chapter.

18. Apply materials to all remaining models. Try adding multiple materials to one or two of them.

19. **Save** all files before moving on.

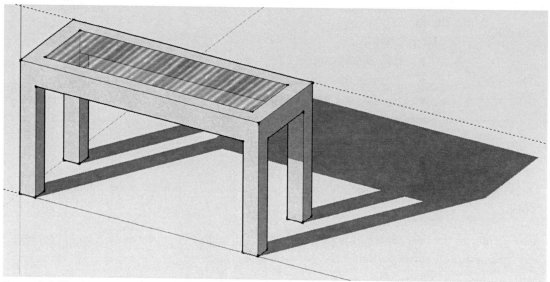

Figure 9.6 Shadows turned on via the View menu

Creating Custom Materials

Creating a custom *Material* is fairly easy in SketchUp. The most difficult part is finding, or creating, a good repeating image; these are images which are seamless and don't reveal an edge when painted on a surface (Figure 9.7). There are a number of sources for tileable (or repeating) images online. Some are free and some are not. Also, if you have other graphics, CAD or rendering packages such as Autodesk Revit or AutoCAD installed on your computer (which is common in a school lab), you have access to a number of images which are installed with that software. The Autodesk textures, for example, can be found here: C:\Program Files (x86)\Common Files\Autodesk Shared\Materials\Textures.

Figure 9.7 Example of repeating image

The next few steps will walk you through the process of creating a new material. We will look at a real-world example by downloading an image file from a carpet manufacturer. This image will then be used by the new material, which can then be painted on any surface.

20. Open the **Materials** window if necessary.

21. Click the **Create Material** icon in the upper right (Figure 9.8).

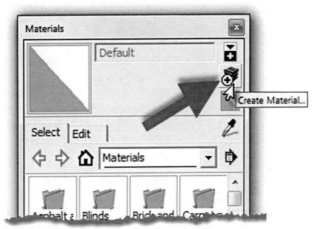

FIGURE 9.8 Create a new material icon

You are now presented with the *Create Material* dialog shown in Figure 9.9. First you will give the material a name; if you don't, the default name provided will be used, which is *Material1* in this example.

A material consists of either a **solid color** or a **texture**, i.e. a raster image file. There is a third option, which is a combination of the color and texture, when *Opacity* less than 100% is specified. This option will show some of the color through the texture depending on the *Opacity* setting.

All of these options are pretty straightforward. One thing to note about selecting a color is the ***Picker*** drop-down. Here you can change from selecting a color visually to entering specific values. This is great when you are, for example, specifying paint from a manufacturer who provides the **RGB** values for all their colors (e.g., *Sherwin-Williams*).

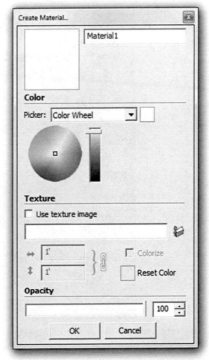

FIGURE 9.9 Create material dialog

Next you will use your internet browser to access the *Shaw* website.

22. Browse to the following URL:

    http://www.shawcontractgroup.com/ProductSpec/Show/59483

Notice that the last number is the model number for the *Shaw* carpet we are using. This can be changed to another known model number to quickly access similar information for another *Shaw* product.

23. Follow the steps shown in Figure 9.10 to download an image file for the specified carpet; select a location on your hard drive to save the file.

FIGURE 9.10 Downloading Shaw carpet image file

If you open the file you downloaded it should look similar to what is shown in Figure 9.11. Notice the text across the bottom gives a disclaimer and information about the carpet in the image. In our case we need to crop this image down from a 9x9 tile pattern to 8x8 to get rid of the text. We will select our pattern from the upper left corner to get a pattern which will repeat properly in SketchUp.

24. Open the image file in an image editing program such as Adobe Photoshop and crop the image down as shown in Figure 9.12.

FIGURE 9.11 Downloaded image file

Next you will switch back to SketchUp and associate the new image with your material.

FIGURE 9.12 Cropped Shaw carpet image file

Now you will modify the new material to represent a tile carpet from the *Shaw Flooring* image you just downloaded.

25. Change the name to **Shaw Carpet**.

26. Check the **Use texture image** option (Figure 9.13).

As soon as you do this, a *Choose Image* dialog will appear. If you need to select another image later, you can click the icon to the right of the image file name.

27. **Browse** to the carpet image you just downloaded and cropped.

    a. If you do not have this file, you can download it from the internet at SDCpublications.com.

    b. Once you have selected the image, click **Open**.

We happen to know that these carpet tiles are two foot square, so eight tiles would be sixteen feet. We need to tell SketchUp our image, which represents an 8 tile x 8 tile area, is sixteen feet in each direction.

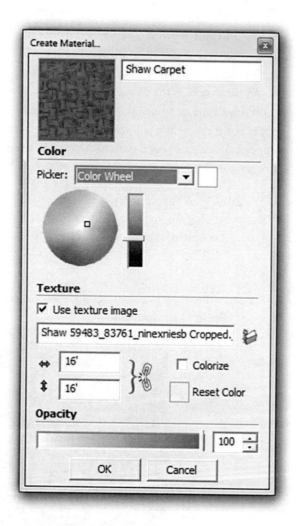

FIGURE 9.13 Custom material creation

28. Change the *Width* and *Height* to **16′** as shown in Figure 9.13.

Click **OK** to finish creating the new material.

29. In a new file, sketch a **rectangle** on the floor. Make it large enough to show 3 or 4 tiles in each direction.

30. **Paint** the face with your new material.

You should now see the carpet squares in the model. The image to the right shows a larger area covered with the carpet tiles. This is fairly realistic because the tiles are often rotated when they are installed. So downloading a larger sample area, as we were able to do, really helped. Not all product manufacturers provide this option. This would also include stone or ceramic floor and wall tiles, ceiling tiles and more.

Sometimes you need to download, scan or photograph what you can and then use an image editing program to make a workable image. In our carpet example, if you could only download one 2′x2′ image, you could create a larger image, in Photoshop for example, and copy plus rotate to create a similar result. For some materials, such as tile, it is a little trickier as you have to make sure the grout lines work when tiled.

Knowing how to use a good image editing program really helps!

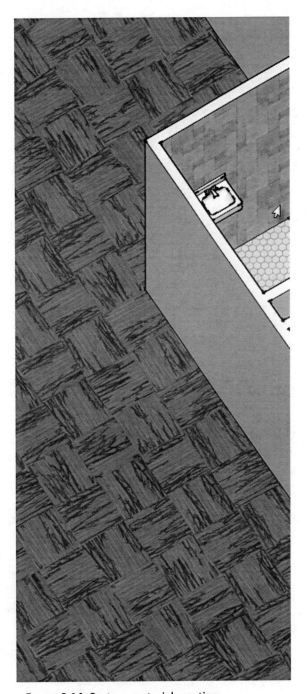

FIGURE 9.14 Custom material creation

Calculating the Area of a Material

SketchUp has the ability to report the area for a material used in the model. Follow these simple steps to derive the square feet of any material:

- Open the **Materials** window
- Switch to view the **In Model** materials
- **Right-click** on the desired material (see Figure 9.17)
- Select **Area**

You will the see the result as shown in Figure 9.15 below. If you plan on using this information, say to estimate material cost, you will want to use extra care when placing, i.e., painting, the material in your model. Applying materials by layer or to multiple faces at once might allow "extra" faces to be included in the final number.

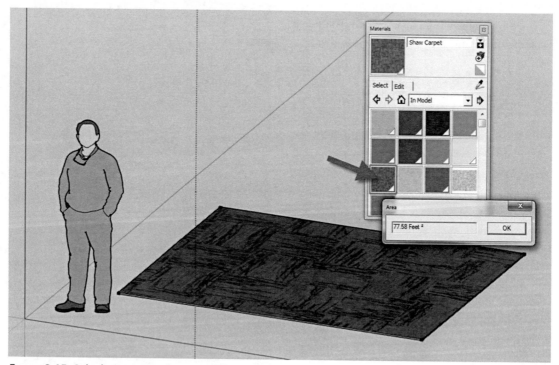

FIGURE 9.15 Calculating area of a material

Once a custom texture has been assigned to a material, the texture representation is saved in the SketchUp model. You <u>do not</u> have to maintain the original texture file with the same name and in the same location on your computer's hard drive.

Repositioning a Material

Occasionally you will want to move, rotate or change the scale of a material.

When you right-click on a face that has a texture (i.e. a raster image) applied to it, you have a few special options available in the pop-up menu (Figure 9.16).

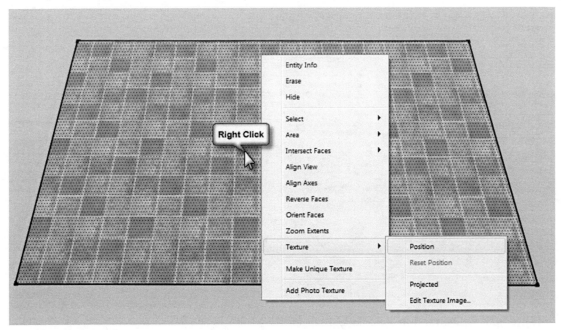

FIGURE 9.16 Right-clicking on face with texture-based material applied

Selecting **Texture → Position** from the pop-up menu places you in an edit mode where you may scale and/or rotate the material (Figure 9.17).

Clicking and dragging on one of the four colored icons initiates the following adjustments to the texture:

- **Red icon:** Moves the texture on the face
- **Green icon:** Scales and/or Rotates the texture on the face
- **Yellow icon:** Distorts texture
- **Blue icon:** Scale/Sheer the texture

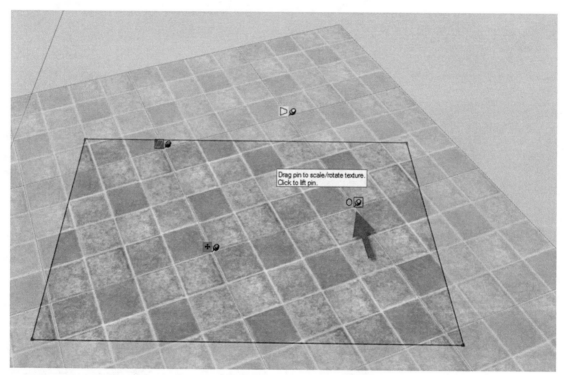

FIGURE 9.17 Position texture edit mode: Dragging the scale/rotate icon

Clicking outside of the visible extents of the material finishes the edit mode. These adjustments only affect this one face. Some of these changes, such as scale and rotation, can be made on a project-wide scale by editing the material (via the Material window).

The pushpins next to each color**Error! Bookmark not defined.**ed icon can be unpinned to repositioned when needed. This is helpful when you know a texture is a certain dimension between two specific potions. Placing the pushpins at these points on the texture allows you to more easily snap those points to corners or intersections in the SketchUp model.

Creating a Custom Collection

When you initially create a material, it only exists in the current model. If you want to access your custom materials in other SketchUp models you will have to create a special location to store them called a *collection*. A collection is basically a folder on your hard drive where you save each material as an individual file with the extension SKM.

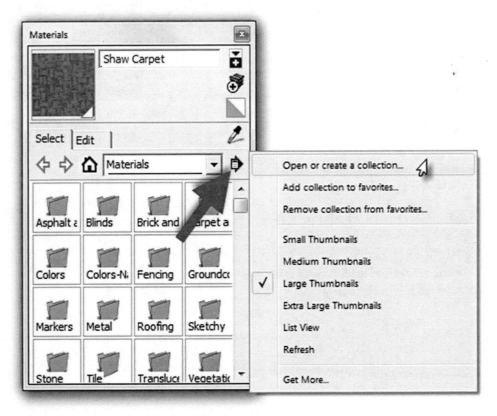

FIGURE 9.16 Creating a material collection

Before you point SketchUp to a place to store your custom collections on your computer, you need to create a custom folder. This way you can keep your materials separate from other files on your computer.

31. Using *Windows Explorer*, create a folder called **Custom Materials**.

    FYI: *You can create this folder anywhere on your computer. If these materials need to be shared with others, in your class or office, you can create this folder on a server. You could also save it in the Cloud; maybe in Dropbox, for example.*

32. Click the **Details** icon, which is pointed out in Figure 9.16.

33. Select the **Open or create a collection** option from the fly-out menu.

34. **Browse** to the folder you create in Step 31 and then click **OK**.

Now you need to save one of your *In Model* materials to the collection folder.

35. Switch to the **In Model** materials.

36. **Right-click** on the "Shaw Carpet" material and select **Save As** from the pop-up menu (Figure 9.17).

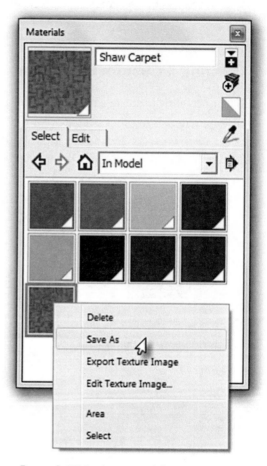

FIGURE 9.17 Saving material

37. **Browse** to the *Custom Materials* folder and click **OK**.

You can now see the custom material, and any others, from any SketchUp model, which have been saved there.

If you want to see these materials all the time, you need to add it to the favorites list. Here is how you do that:

38. Select the **Details** icon again, and then select **Add collection to favorites.**

39. Browse to the *Custom Materials* folder again, and then click **OK.**

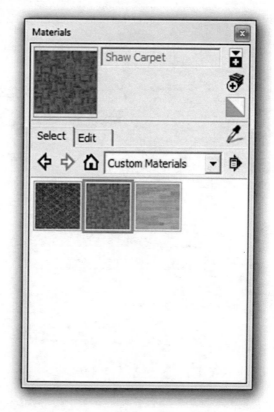

FIGURE 9.18 Saving material

You now have the basics of how to use and manage materials in SketchUp.

Sample project courtesy of LHB, Inc. and Hybird Medical Animation

Chapter 9 Review Questions

The following questions may be assigned by your instructor as a way to assess your knowledge of this section. Your instructor has the answers to the review questions.

1.   Materials cannot be rotated. (T/F)

2.   The "face style" needs to be *Shaded with Textures* to see the materials. (T/F)

3.   The only way to add a material is to click on every surface. (T/F)

4.   Once a surface has been extruded, it cannot be modified. (T/F)

5.   Erasing an edge (i.e. line) between two coplanar surfaces results in the two surfaces becoming one. (T/F)

6.   You cannot create custom materials. (T/F)

7.   If you want to access your custom materials in other SketchUp models you will

      have to create a custom _____ .

8.   Right-clicking on a face gives you some powerful selection options. (T/F)

Chapter 10
Detailed SketchUp Model

It may be helpful at times to create a more detailed SketchUp model. Some do this for an entire project; others just use it to focus on small custom portions of the project. This section will focus on the latter, and walk through the process of modeling a reception desk. This could also be done for built-in casework, such as cabinets or bookshelves, custom furniture, etc.

You will learn a few new concepts here, such as how to cut a section through your model. However, you already know most of the steps necessary to create the reception desk model. This tutorial will just help you learn to think about what you are trying to model and the various ways in which you can approach the problem given SketchUp's toolset.

For this tutorial you will start a new SketchUp model. You could work within a project model if you wanted. However, for something detailed like this reception desk, it is easier to work in a separate model and then load it into the project.

1. Start a new SketchUp model.

2. Create a rectangular box (Figure 10.1):

    a. Sketch a 10'-0" x 3'-0" rectangle on the ground/floor.

    b. Use ***Push/Pull*** to extrude the rectangle 3'-6" upward.

The rectangle represents the outer extents of what we think the desk will be. The 3'-6" high is based on the desire to have a 42" high transaction surface; this is a surface for customers to write on and look at paperwork provided by the receptionist. The final desk may extend outside of this box. But, we will use this as a starting point. Think of it as a piece of wood you will carve portions away from.

The default person can remain and be used as a scale reference.

FIGURE 10.1 Start with a rectangular box

3. Sketch an outline as shown in Figure 10.2):

    a. The three dimensions shown are:

        i. **1'-0″** – transaction surface (1½″ thick)

        ii. **2'-6″** – typical desk height

        iii. **6″** – thickness of wall supporting transaction surface.

        iv. See Figure 10.3 for intended goal

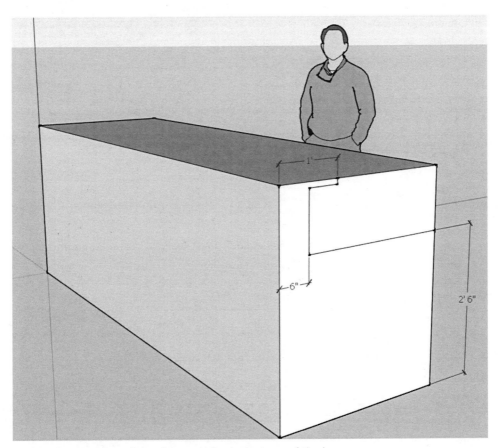

FIGURE 10.2 Sketch outline of areas to be carved out of the box

4. Use the ***Push/Pull*** tool to carve out a portion of the box behind the transaction counter (Figure 10.3):

    a. Pull the surface until it snaps to the opposite end of the box so the section being pulled completely disappears once you let go of the mouse button.

This step also defined the height of the work surface (30″ high).

The next step will be to carve out a portion of the box to define the knee space below the work surface. This often requires brackets to support the work surface, but that will be ignored here – we will save that for the *Construction Documents* phase.

5. Sketch the **lines** shown in Figure 10.4 on the back face of the desk:

    a. 1½″ thick work surface

    b. 6″ side walls

FIGURE 10.3 Using Push/Pull to remove a portion of the box

6. Use ***Push/Pull*** to "push" the face back 2′-6″.

This modification leaves a 6″ wall to support the transaction counter (3′-0″ − 2′-6″).

FIGURE 10.4
Sketch outline for open
space below work surface

FIGURE 10.5 Use Push/Pull to carve out leg space

7. Using similar steps, define a 3″ deep and 4″ tall **toe space** on the front side (or public side) of the reception desk (Figure 10.6).

8. Sketch the **lines**, equally spacing the vertical lines, as shown in Figure 10.6:

    a. Draw one centered and then copy that line 2′-6″ each way

    b. Copy via the *Move* tool while holding down the Ctrl key

FIGURE 10.6 Toe space and line work added

The lines sketched in the previous step will be used to define metal panels. If the desire were to have a more exaggerated reveal, you could sketch the reveal profile and then use *Push/Pull* to create it. However, in most cases a line can adequately suggest a reveal without the extra work.

The next step is to start adding materials. You will have to use *Orbit* to apply materials to all sides of the desk.

9. Add the horizontal line at each side, aligned with the top of the top space (Figure 10.7). This will define the base material.

10. Use the *Paint* tool to add **Metal Aluminum Anodized** to the front and sides of the desk (Figure 10.7).

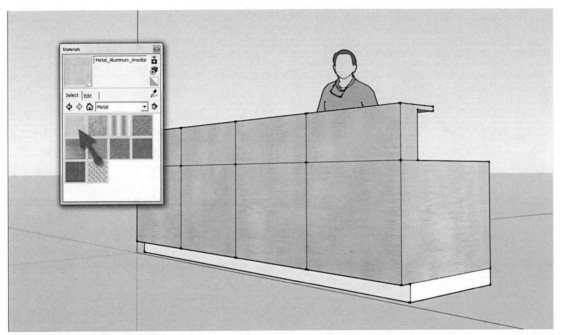

FIGURE 10.7 Adding materials

11. Sketch additional *lines* and use *Push/Pull* to refine the transaction surface as shown in Figure 10.8:

   a. Make the front edge extend 1″ out

   b. Make the sides extend ½″ out

12. Add more materials using the **Paint** tool:

    c.   Markers\Yellow Green – work surfaces

    d.   Tile\Tile Ceramic Natural – base

FIGURE 10.8 Defining transaction counter

Next you will be adding decorative fasteners at the corner of the panels. To help locate them, you will add guides and then use the intersection of these guides to locate the fastener. The fastener will be created as a *Component* so they can all be changed at once, if need be later.

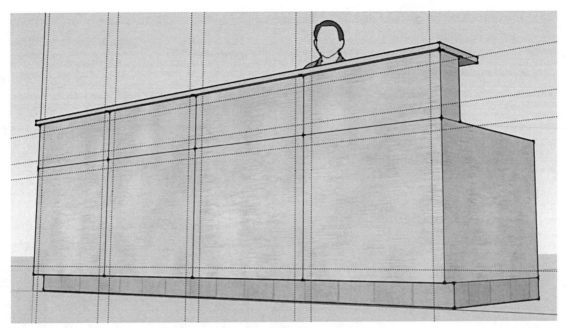

FIGURE 10.9 Adding guides to define decorative fastener location

13. Add **Guides** 2″ from the edge of the panels as shown in Figure 10.10.

   REMINDER: *Use the* Tape Measure *tool.* **TIP:** *After you create the first gridline, the offset of 2″ will be the default—meaning SketchUp will snap to the location.*

14. Make a **2″** diameter and ½″ thick disk.

   e. Create the first fastener at the intersection of two gridlines as shown in Figure 10.10.

   f. **Paint** with the same material as the work surface; *Marker: Yellow Green.*

15. Once finished, select disk (all edges and faces), right-click and select **Make Component**. Name it: **Decorative Fastener**.

16. **Copy** the fastener *Component* around and then erase the guidelines (Figure 10.11).

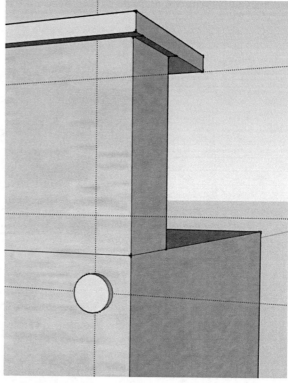

FIGURE 10.10 Create fastening and turn into component

FIGURE 10.11 Decorative fasteners added

It is often helpful to use the ***Section Plane*** tool to look at your model in section. This can be done vertically or horizontally. In this exercise you will look at a vertical application. This could also be applied horizontally to a floor plan (i.e., top) view of a more detailed floor plan model.

17. Select **Tools → Section Plane** from the menu.

18. Hover your cursor over the side of the desk, until the *Section Plane* appears as shown in Figure 10.12, and then click.

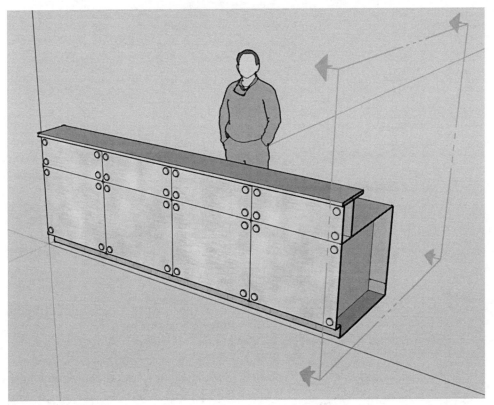

FIGURE 10.12 Adding section plane

19. Select the *Section Plane*, and then use the **Move** tool to reposition it so you see the desk in section as shown in Figure 10.13.

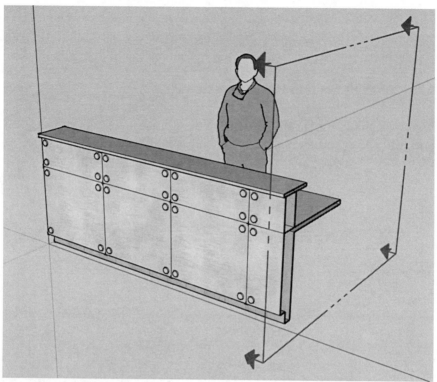

FIGURE 10.13 Adjusting section plane location

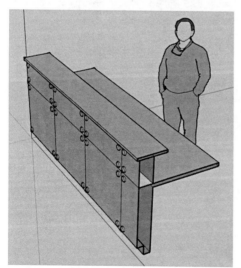

FIGURE 10.14 Section plane visibility turned off

20. From the *View* menu, toggle off **Section Planes**.

Notice how the *Section Plane* element is no longer visible but the model is still in section. This can create a nice presentation drawing. Additionally, toggling off the **Section Cuts** item in the *View* menu restores the entire model.

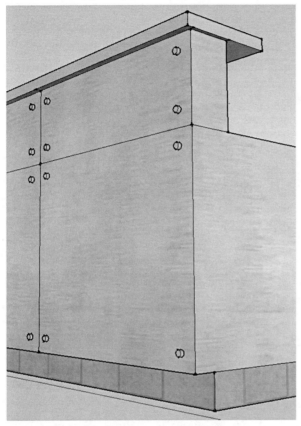

FIGURE 10.15 Edit fastener component

It is decided that the 2″ diameter fasteners do not look good aesthetically. Therefore, you decide to change them to 1″ diameter. Changing one *Component* changes them all!

21. Toggle off **Section Cuts** from the *View* menu

22. Edit the *Component*; Pick the *Select* tool and then double-click on it.

23. Use the **Scale** tool from the *Tools* menu.
    a. Select all the geometry
    b. Select the **Scale** tool
    c. Click on of the center-corner grips
    d. Hold **Ctrl+Shift**
    e. Drag and release the mouse
    f. Type **.5** and **Enter**

Another option, rather than copy, is to delete the edges and faces within the *Component* and create the geometry from scratch. When finished with the *Component* edit, all instances will be updated.

This last step will cover placing furniture in your model.

24. Select **Window → Components**.

25. Type **Haworth** and press **Enter**.

26. Select one of the chairs listed, that are created by *Haworth*. Place it in your model (Figure 10.16).

If you open the **Layers** dialog, you will see new *Layers* created for the chair. The arm rests have their own layer. Therefore, the arms can be turned off if needed.

FIGURE 10.16 Placing components

It is good to remember that some manufacturers have created SketchUp content and made it available on the internet. This can save a lot of time and help to quickly make your models look really great!

This model can now be printed and added to a presentation board. You can use File → Export → 2D Graphic to create a raster image file on your hard drive for use in Adobe Photoshop or InDesign and Microsoft Word. This export feature prompts you for a file name and location.

Chapter 10 Review Questions

The following questions may be assigned by your instructor as a way to assess your knowledge of this chapter. Your instructor has the answers to the review questions.

1. To model this reception desk you must work in a new, empty file. (T/F)

2. The section tool permanently deleted things once they are not visible on the screen. (T/F)

3. Components make it easier to quickly adjust an element used in multiple locations. (T/F)

4. Simple 2D lines were used to represent panel joints. (T/F)

5. The offset distance used becomes the default until changed. (T/F)

6. How high did you make the transaction counter? _____

7. Some manufacturers have created SketchUp content and make it available on the internet. (T/F)

8. Layers are never created when placing Components. (T/F)

Chapter 11
Working with Styles

In addition to SketchUp's ease of use, another great attraction for a designer is the aesthetic quality of the images which can be produced. With very little effort, images can be created with "styles" ranging from technical to hand drawn. This can affect all aspects of the model, including the background. As you will soon discover, you have a number of options to control the graphic styles.

The template you started from automatically has a few *Style* related items set (Figure 11.1). For example, the **Edge Extensions** feature causes the line ends to extend and overlap slightly to accentuate the beginning and end of each line. This is something many designers do when sketching by hand. Another setting from the typical architectural template is **Profile**. This feature thickens lines at outside edges and plane changes to help with edge and depth perception. We will explore how to control these options and more in this chapter.

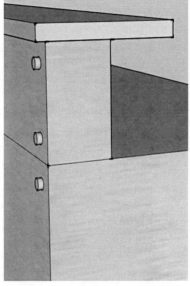

FIGURE 11.1 Default style settings

There are three main ways in which we can work with *Styles* in SketchUp.

- View menu toggles
- Styles toolbar
- Styles browser

We will use the reception desk model created in the previous chapter as the "sandbox" to experiment with the *Style* tools. Maybe we are preparing presentation images of this new desk design for a client.

1. Open your **Reception Desk.skp** model.

Next you will create a copy of the file so you don't have to worry about messing up the original.

2. From the **File** menu, select **Save As**.

3. Name the new file: **Reception Desk – Styles.skp**.

View Menu Toggles

First we will take a look at the toggles related to *Styles*. These allow you to control which *Style* settings are applied to the model, if any.

First we will turn off *Profiles*.

4. Select View → Edge Style → **Profiles** (Figure 11.2).

Notice all the lines, or edges, are now the same thickness. Next you will turn off all the *Edge Styles* to see the stripped-down basic model.

5. From the *View* menu, uncheck all the options under **Edge Style**.

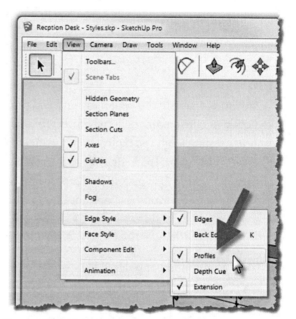

FIGURE 11.2 Style toggles

Notice the model is very simple looking, and does not have much definition (Figure 11.3). The only style still applied now is the *Face Style*: **Shaded with Textures**. This toggle can also be seen from the *View* menu (Figure 11.4). At least one *Face Style* must be selected, so this is a toggle between options shown in the menu fly-out. Don't change these settings back yet.

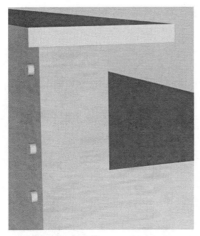

FIGURE 11.3 Result of turning off all Edge Styles

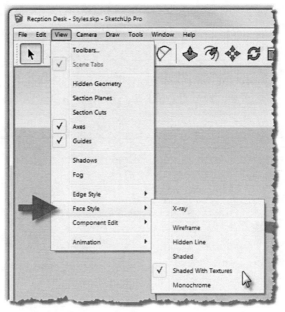

FIGURE 11.4 Face Style toggles

Styles Toolbar

SketchUp provides a *Styles* toolbar which allows the designer the opportunity to quickly adjust the active *Style*. The first thing you need to do is make sure the toolbar is turned on.

6.   To turn on the toolbar, select **View → Toolbars → Styles**.

You now have a toolbar open, which can be docked (i.e., attached to the top edge of the screen).

From left to right, these commands are:
X-Ray, Back Edges, Wireframe, Hidden line, Shaded, Shaded with Textures, Monochrome.

The first two icons are toggles which can be on or off in addition to one of the remaining five icons. This toolbar mainly controls the toggle between *Face Styles*. Only the *Back Edges* tool is an *Edge Style* setting.

7. Ignore the first two icons for now, and click on the last five icons to see what each one does. You should see results similar to those shown below.

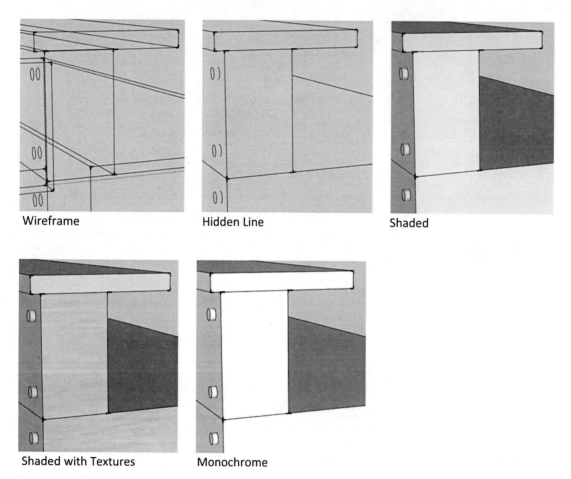

Wireframe   Hidden Line   Shaded

Shaded with Textures   Monochrome

As soon as you click on *Wireframe* or *Hidden Line* you should notice that *Edges* (an *Edge Style*) are toggled back on. Those *Face Styles* require *Edges* be on.

Next you will review the effect that *X-Ray* and *Back Edges* have on the model – these are the first two icons on the *Styles* toolbar. For these two icons, they can both be "off" or just one of them can be "on".

8. With *Monochrome* still selected, click on the **X-Ray** icon, the first icon on the *Styles* toolbar.

As the name implies, you can see through everything, but without the radiation exposure (Figure 11.5). This allows you to get a better look at your model without the need to be continually orbiting around the model. This is usually used more as a "working" mode rather than a presentation technique. That's not to say there is not a situation in which this would be a great presentation style.

Take note that the **X-Ray** icon looks like it is pushed in, to indicate it is active. Let's switch to *Back Edges* next, and see what this does to our model.

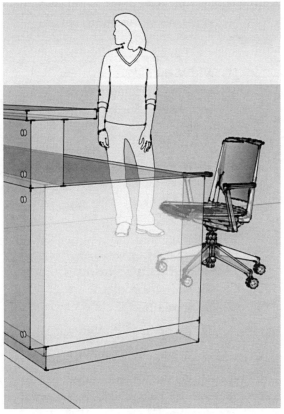

FIGURE 11.5 X-Ray mode toggled on

9. With *X-Ray* still selected, click on the **Back Edges** icon on the *Styles* toolbar.

Notice how the *X-Ray* mode is forced "off" and the *Back Edges* mode is now "on". This mode makes all edges, which are otherwise hidden on faces based on the direction you are viewing the model from, dashed. Put another way, this mode makes all "back edges" visible and dashed.

The image to the right (Figure 11.6) shows the back edges dashed for our reception desk.

10. With *Back Edges* active, try clicking on each of the five icons on the right side of the *Styles* toolbar.

You should have noticed, as soon as you selected *Wireframe*, the first two icons become grayed out because all edges are showing in this mode.

11. Before moving on, make sure **Shaded with Textures** is selected plus *X-Ray* and *Back Edges* are off.

Everything you have done with *Styles* thus far has been an override to whatever the current *Style* is for your model. Next you will look at how to apply, manage and create custom *Styles*.

FIGURE 11.6 Back Edges mode toggled on

Styles Browser

Now we will dig in and look at some really cool things we can do with *Styles*. Keep in mind that *Styles* do not actually change your model in any way. They just change how the edges and faces are presented, if at all. In this section we will look at how we can apply pre-created *Styles* and edit the current *Style*.

The first thing we need to do is open the *Styles Browser*, or dialog box.

12. Select **Window → Styles**.

You now see the *Styles Browser* shown to the right (Figure 11.7). At the top you see a preview of the current style, the style name and a description. Below that are three tabs where you can **Select** (i.e., apply) a pre-baked *Style*, **Edit** the current *Style* and **Mix** various parts of the pre-baked *Styles* to get new results.

If you click on the title bar for this browser, it will collapse to maximize the visible portion of the canvas area (Figure 11.8). Clicking the title bar again expands it. You may close it by clicking the red X in the browser. You can also drag it to a second monitor if you have one.

We will start by applying pre-baked *Styles* to get some quick and amazing results.

13. On the **Select** tab, open **Assorted Styles**.

   a. This can be done via the drop-down list directly below the tabs, or by clicking the "folder" image/icon in the preview panel area below (Figure 11.7).

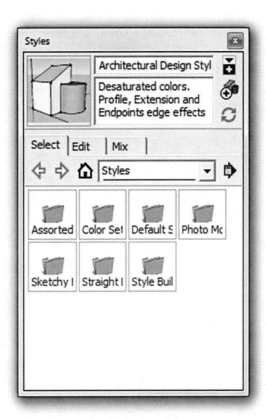

FIGURE 11.7 Styles browser

FIGURE 11.8 Styles browser collapsed

Notice, in Figure 11.9, that the drop-down list indicates the current collection/folder you are in and the preview panel below shows several previews. You will try several of these on you model.

Just before proceeding with the next step, take note that the *Section Cut* and *Guides* are currently off. You turned these off via the *View* menu previously in this chapter. These features will become visible again because these settings are saved in the *Styles*.

14. In the *Assorted Styles* folder, select the second item (pointed out in Figure 11.9). If you hover over the preview before clicking, you will see the *Style* name: **Brush Strokes on Canvas**.

Notice the overall aesthetics of the entire model exhibition have changed drastically. As you will see, *Styles* control much more than you have seen so far. Next, you will turn off the *Section Cut* and *Guides* to clean up the image.

15. From the *View* menu, turn off the following:

    a.   Section Planes

    b.   Section Cuts

    c.   Guides

FIGURE 11.9 Styles browser collapsed

Your model should now be presented as shown in Figure 11.10 on the next page. With just a couple clicks you have made a very compelling design presentation.

16. In the *Style Browser*, click on the same *Style* again: **Brush Strokes on Canvas**.

Notice the *Section Cut* and *Guides* came back. The changes made in Step 15 did not change the saved version of the *Style*. More on this in a moment.

Figure 11.10 Brush Strokes on Canvas applied to model

17. Click on **all** the *Styles* listed under *Assorted Styles* to see how they look on your reception desk model.

Notice that some styles change the background so that the horizon line is not shown and the outer edges of the screen are redefined.

In the example to the left, Figure 11.11, the *Style* **Pencil on Tracing Paper** applies a very whimsical line to the model edges. This makes the model look more hand drawn. This is great when you want to convey an initial design concept to a client, but don't want the model to look so polished and complete that they are afraid to suggest a change.

18. Set the current *Style* to **Pencil on Tracing Paper**.

19. Turn off the *Section* and *Guide* features again.

FIGURE 11.11 Pencil on Tracing Paper applied to model

Notice, in the *Styles Browser*, that the preview in the upper left has two circular arrows superimposed on it (Figure 11.12). This indicates that the current *Style* has been overridden. You may either leave things as they are or click the update button to save the changes. If you don't save the changes, it will take more time to get back to this "style" if you, for example, try another *Style*.

In the next step you will update the *Style*. The one thing to be aware of is that you are only updating the COPY of the *Style* in your current model. This makes sense as you would probably not want to change a default, out of the box, *Style* that comes with the software for all future models.

20. Click the **Update Style** button pointed out in Figure 11-12.

Pencil on Tracing Paper is now updated, and when clicked on in the future, the *Section Cut* and *Guides* will not be visible. However, you have to click on the right version of the *Style…*

If you click on the *Pencil on Tracing Paper Style* shown to the right, under *Assorted Styles*, the *Guides* and *Section Cut* will actually appear.

21. Click on the *Style* **Pencil on Tracing Paper** again, under *Assorted Styles*.

You should see the *Guides* and *Section Cut*. This is the version of the *Style* which comes with the software. Anytime you apply a *Style*, a COPY of it is saved in your model. You will see this in the next step.

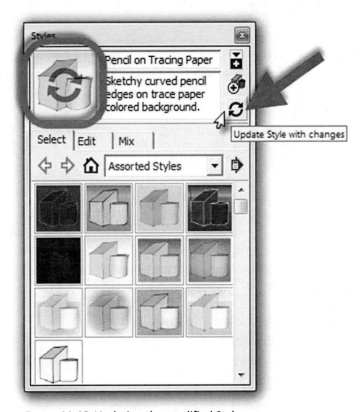

FIGURE 11.12 Updating the modified Style

22. Change the Collection/Folder from *Assorted Styles* to **In Model** from the drop-down list in the *Styles Browser* (Figure 11.13).

You should now see all the *Styles* you have applied (i.e., tried in this case) to your model (Figure 11.13). Just applying a *Style* one time, for just a moment, copies it into your model.

While several extra *Styles* in your model probably will not "break the bank" in terms of file size, it may be prudent to "clean house" and delete the extra unneeded *Styles*.

To remove a *Style*, simply right-click on the preview and select *Delete* from the pop-up menu.

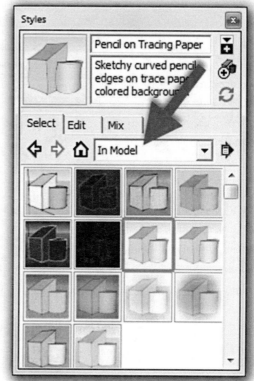

FIGURE 11.13 In Model styles listed

23. Try several more *Styles* from other "collections" before moving on.

24. When finished, apply your copy of **Architectural Design Style** from the *In Model* area.

   a. This *Style* can also be found in the *Default Styles* collection.

   b. Turn off the *Section Cut* and *Guides* if they are on.

The next two images show two examples from the collection *Style Builder Competition Winners*. These styles have some pretty amazing techniques applied.

Figure 11.14 Style Builder Competition Winners: Pencil Sketch with Darker Traced Lines

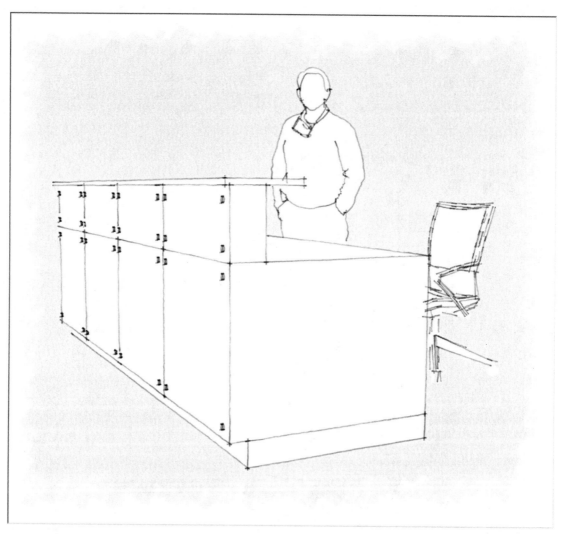

FIGURE 11.15 Style Builder Competition Winners: Pencil Edges with Whiteout Border

Styles Browser: Edit Tab

To wrap up this chapter on *Styles*, you will take a look at the **Edit** tab in the *Styles Browser*. This is where you can control ALL the options related to how your model appears on the screen and prints.

25. If not already open, open **Styles** (aka *Style Browser*) from the *Window* menu.

26. Click on the **Edit** tab.

You should see something similar to Figure 11.16. Notice the FIVE icons highlighted by the added arrow. Clicking on each of these icons reveals different settings in the [windowed] area below. Currently in this image, the fifth, or last, icon is selected. You can tell, because the icon/button appears pushed in and the label to the right of the buttons changes depending on which one is selected.

From left to right, the icons are:

- Edge Settings
- Face Settings
- Background Settings
- Watermark Settings
- Modeling Settings

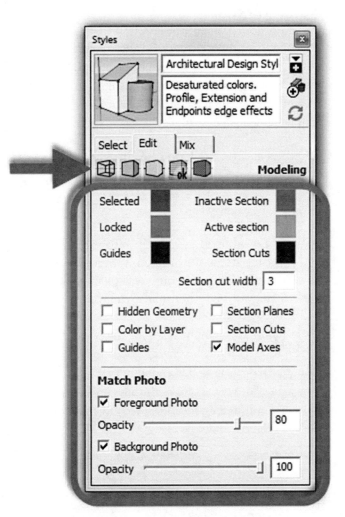

FIGURE 11.16 Style Browser; Edit tab

Edge Settings

Keep in mind, any changes made here directly affect the current Style listed above – and your model. Next you will make a few changes to the default Architectural Design Style to see what can be done here.

> 27. Click on the **Edge Settings** icon (Figure 11.17).

Notice the options change below.

> 28. Try adjusting the settings to see how they affect your model. When finished, make your settings match Figure 11.17 before moving on.

Your model should now look similar to Figure 11.19, shown on the next page. Notice the lines continue past their natural endpoint to exaggerate the corners. Also, the lines appear thicker closer to the viewer (i.e., you) to convey a sense of depth.

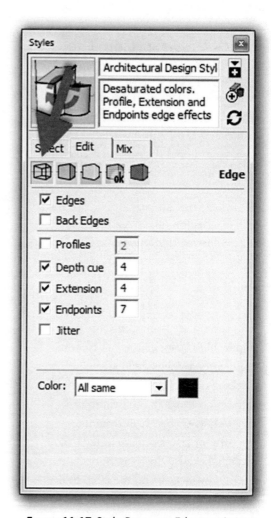

FIGURE 11.17 Style Browser; Edge settings

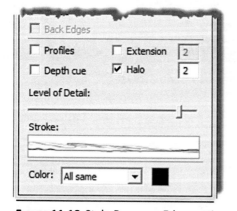

FIGURE 11.18 Style Browser; Edge settings

Some *Styles* utilize a newer feature called *Non-Photorealistic Rendering (NPR)* which presents different options (Figure 11.18). This feature allows you to scan in your own custom hand sketched line drawing and use it for your edges.

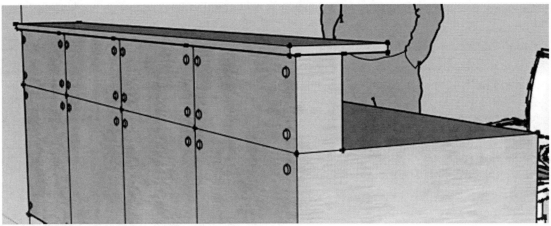

FIGURE 11.19 Changes to the edge settings

Face Settings

29. In the *Styles Browser*, under the **Edit** tab, Click on the **Face Settings** icon.

You only have a few options here and you have already looked at several of them via the *Styles* toolbar (Figure 11.20).

An important concept to understand is that each face has two sides. Each side can receive a different material. When a face is initially created, the "front" side has a color applied, as does the "back" side. Those default colors are specified here. You can use the **Display shaded using all same** icon to verify face orientation. If a face is backwards, just select it, right-click and select *Reverse Faces*.

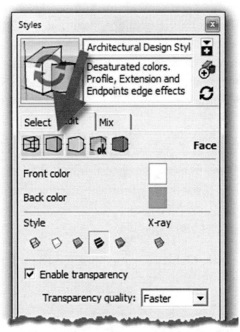

FIGURE 11.20 Style Browser; Face settings

If you are modeling a grate floor, and want to see through it to the structure below, you have to apply the transparent grate material to both sides of the face (Figure 11.21).

Another way in which this concept could be leveraged is in visualizing small spaces. In the space, you would apply the desired material. On the back side of the faces of the space, apply a transparent material. This allows you to view the space from outside the space – seeing through only the walls directly in front of you!

FIGURE 11.21 Transparent material applied to both sides of face

The last thing to mention about *Face Settings* is the transparency options. In extremely large models you probably want to turn this off while working on the model. Otherwise the display can be very slow when panning and orbiting the model. The same applies to the quality setting. You might bump this up when printing or exporting images. Otherwise, keep it set at "faster" for optimal performance.

30. Click the **Display shaded using all same** option (hover cursor for tool tip).

All the faces in your model should be white, which represents what SketchUp considers the "front" face. If you see any light blue faces, you will want to fix them. Keep in mind, you have to orbit the model and look at each face somewhat straight-on; otherwise you might see a surface that is shaded and just looks light blue.

31. **Orbit** your model and correct any backwards faces as needed.

   a. Select the light blue face

   b. Right-click

   c. Select **Reverse Faces** from the pop-up menu

32. In the *Styles Browser*, switch back to **Display shaded using textures**.

Next we will take a look at the background settings.

Background Settings

33. In the *Styles Browser*, under the **Edit** tab, click on the **Background Settings** icon.

Here you see three color swatches (Figure 11.22): one for the *Background*, one for the *Sky* and another for the *Ground*.

FIGURE 11.22 Style Browser: Background settings

Sometimes you just want the focus of your model to be the design, not the background or the horizon line, especially seeing as you never really see the horizon as a straight line unless you are looking at the ocean. In this case, you can uncheck (i.e., turn off) *Sky* and *Ground* and only see a solid color in the ***Background***.

As soon as you turn on the ***Sky***, you will see the horizon line even though the ground plane is not turned on. This is because the sky stops at the horizon. If you are looking up, you will only see *Sky*; if you are looking down, you would only see *Ground* or *Background*.

If you turn on the ***Ground*** you will not see the *Background* color unless the *Ground* plane is set to be transparent via the slider shown in Figure 11.22. You might want the *Ground* to be transparent if you need to show something below, such as structure.

In the world of interior design, you will likely want a solid colored background seeing as we are not typically dealing with the exterior of the building. You will make this change in the next step.

34. **Uncheck** *Sky* and *Ground*, if not already unchecked.

35. Click on the color swatch for *Background*.

36. Set the color to an **off white** (or whatever you want) via the *Color Chooser* (Figure 11.23).

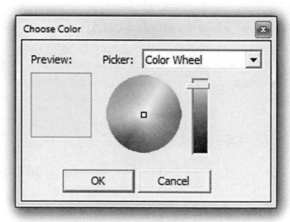

FIGURE 11.23 Choose Color dialog

Notice, in the *Choose Color* dialog, you can select RGB from the drop-down list in the upper right. This allows you to enter the *Red*, *Green*, and *Blue* numbers for a color. This is handy if you want to pick a color from a paint sample, such as from *Sherwin Williams* which provides the RGB numbers for all of their colors.

Shadows

We will digress for a minute and talk about shadows. Even though you do not have the *Ground* turned on, the shadows will still be cast upon the imaginary floor plane; which corresponds to the X and Y axes (the green and red lines).

37. From the *View* menu, click **Shadows** to turn them on.

Your view should look similar to Figure 11.24. Notice the solid colored background and the shadows on the imaginary ground plane.

The light source for the shadows is the sun, which is typically based on the geographic location of the building. For interior designers, shadows are used to convey depth and reveal form within the model (if used at all). The shadows can be thought of coming from an interior light source, but they are too directional for this to be the case. This is where we can take some artistic liberties with our presentation.

Before moving on, let's take a quick look at the *Shadow* settings.

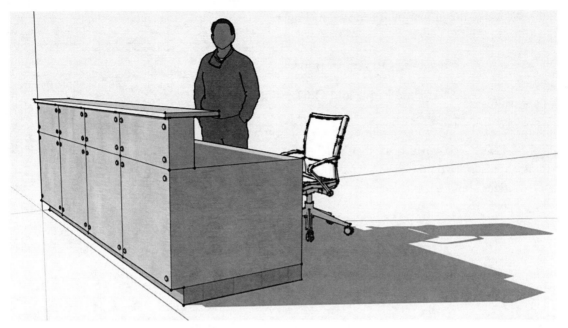

FIGURE 11.24 Solid background color and shadows added

38. From the *Window* menu, select **Shadows**.

The *Shadow Settings* dialog appears. Adjusting the *Time* and *Day* controls where the sun is in the sky, which subsequently determines shadow locations. Notice the check box for "On ground" and "On Faces". Next you will adjust the Light and Dark values to give your model a deeper toned, more realistic feel.

39. Make the adjustments shown in Figure 11.25:

   a. Check the **On faces** option.

   b. Adjust the **Light** and **Dark** values.

Notice the richer color tones with more overall contrast (Figure 11.26). You can also see the shadow of the transaction surface on the work surface because the "On Surface" option was selected.

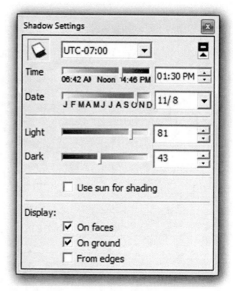

FIGURE 11.25 Shadow settings dialog

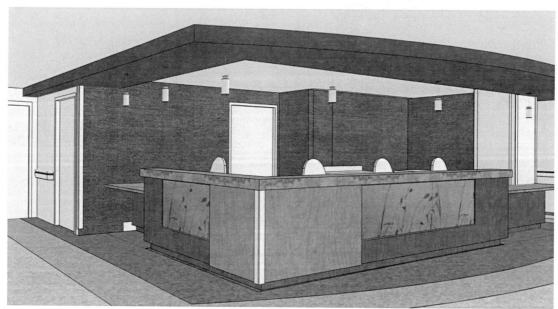

Project Example: Reception Desk.

Image courtesy of LHB; www.LHBcorp.com

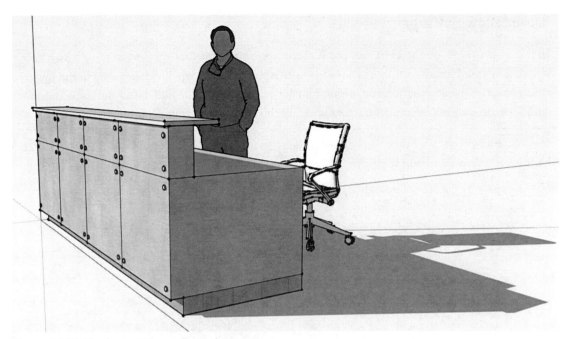

FIGURE 11.26 Shadow settings adjusted

One last thing about shadows… if you want to do **basic daylight** studies in SketchUp, you need to keep a few things in mind. First, your model needs to be fully enclosed. That is, it at least needs to have a ceiling and all the walls to prevent extra light from spilling in. You also need to tell SketchUp where your model is in the world, which can be done via **Window → Model Info → Geo-location**. Finally, you need to make sure your model is positioned according to True North (Solar North). We often use Plan North (or Project North) so all the geometry is orthogonal with the computer screen, which makes things easier to model and place on sheets. Keep in mind that SketchUp does not bounce light around or provide light level values, so this is just a basic investigation of daylight within the model. Now let's get back to talking about *Styles*…

Watermark Settings

In the *Styles Browser*, on the **Edit** tab, the **Watermark Settings** icon exposes the tools which allow you to add a watermark. A watermark can be a company logo which always appears in the lower right corner of the screen, or a background image which can also be faded into the background color. The watermark image can be tiled so it fills the entire background of the given view.

If you use an image of a fabric, you can set the image to be in the background, and fade-in to the background as in the example below.

FIGURE 11.27 Background texture added using the Watermark feature

40. Add a fabric-type background per the following steps:

    a. Locate an image file

        i. Go to **www.pollackassociates.com**

       ii. Browse to a fabric you like

      iii. Right-click on it and select **Save picture as**

      iv. **Save** the file to your hard drive

    b. In the *Style Browser*, click the "+" icon to add a Watermark

        i. Browse to the image you just downloaded

       ii. Select **Background**, and then **Next**

      iii. Adjust the **Blend** to about the middle, and then click **Next**

      iv. Select **Tiled across the screen**

       v. Adjust the **Scale** as desired

      vi. Click **Finish**

Modeling Settings

The *Modeling Settings* control basic things related to the modeling environment. For example, does the *Section Plane* and *Cut* appear when this *Style* is active?

41. Uncheck **Model Axes** (all six boxes should be unchecked).

FIGURE 11.28 Style Browser: Modeling settings

Update Style with Changes

The last thing you want to make sure you do is save all these changes to the *Style*. If you don't save these changes, and then apply a different *Style* temporarily, you will not be able to get back to these exact changes without re-applying each change you made to this *Style*.

42. In the *Style Browser*, click the **Update Style with Changes** icon, which is pointed out in Figure 11.28.

43. **Save** your SketchUp model.

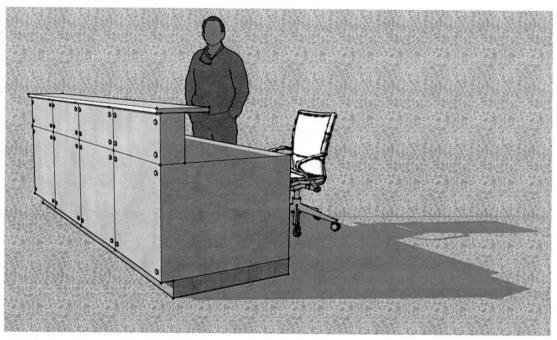

FIGURE 11.29 View with updated style

Chapter 11 Review Questions

The following questions may be assigned by your instructor as a way to assess your knowledge of this chapter. Your instructor has the answers to the review questions.

1. SketchUp provides many predefined Styles. (T/F)

2. List the three main ways you can work with Styles:

    a. _____

    b. _____

    c. _____

3. The Wireframe style hides all faces(T/F)

4. The Style browser is accessed from the Camera menu. (T/F)

5. The style browser can collapse to jus the title bar to minimize it. (T/F)

6. Shadows will appear on the ground even when no surface is present. (T/F)

7. You are not able to select a color for the background. (T/F)

8. Transparency can be turned off in large models to improve performance. (T/F)

Chapter 12
Scenes and Animations

This chapter will teach you how to save different views of your model. You will also learn how to take those saved views and create an animation.

1. Open your **Reception Desk - Styles.skp** model.

Next you will create a copy of the file so you don't have to worry about messing up the original.

2. From the **File** menu, select **Save As**.

3. Name the new file: **Reception Desk – Scenes.skp**.

Creating a Scene

Creating a *Scene* is an essential part of the designer's workflow. This feature allows you to save multiple camera angles plus various options such as the *Style*. The *Scenes* are saved as "tabs" across the top of the *Drawing Window*. A single click on one of these tabs changes the view and, optionally, the *Style*, *Section Cut* and more. These tabs make it easy to jump around the model while you are working on it, or presenting it to a client.

4. Adjust your view of the reception desk to look similar to the last image in the previous chapter – if necessary.

5. From the *View* menu, select **Animation → Add Scene**.

You should now have a new tab, named **Scene 1**, near the top of your screen (Figure 12.1). The blank space to the right is where additional tabs will appear as new *Scenes* are created. Next, you will take a look at the *Scene Manager*, where you can rename the *Scene* to something more meaningful and to see what is saved with it.

6. Right-click on the tab named **Scene 1** (Figure 12.1).

7. Select **Scene Manager** from the pop-up menu.

Notice, in the pop-up menu, a few handy options, such as *Add*, *Update* and *Delete*.

FIGURE 12.1 A scene tab added

8. Click the **Show Details** toggle in the upper right corner (a down-arrow with a plus symbol under it).

9. Change the name to **Desk - Front+Right** (Figure 12.3).

Notice the various options associated with this saved scene (Figure 12.3). The first option determines whether or not this particular *Scene* will be included in an animation (if you decide to create one).

FIGURE 12.2 Scene tab right-click menu

In the *Properties to save* section, a checked item means the settings for that item, for example *Style and Fog*, are saved when the *Scene* is saved, and then re-applied whenever this *Scene* is selected. If the box is not checked, the current values persist.

Next you will reposition the view of the desk such that we are looking more at the front, or what is often called the public side. Once we do this we can then save another *Scene*.

10. Adjust the current view of the model, using **Orbit** and **Pan**, to look similar to Figure 12.4.

11. **Right-click** on the **Scene** tab, and select **Add**.

12. Create additional *Scenes*:

    a. In the order shown, adjust the view per the images shown on the next page.

    b. After each view is achieved, right-click on the last tab and click **Add**.

The new *Scene* is added directly to the right of the selected tab. This will be important if you plan on creating an animation – more on this later.

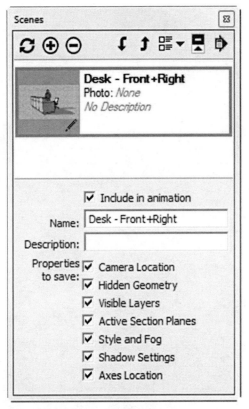

FIGURE 12.3 Scene Manager

FIGURE 12.4 Adjusted view of reception desk

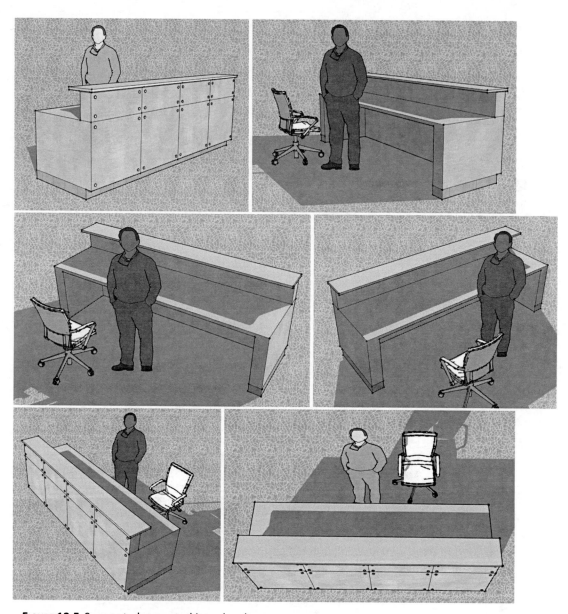

FIGURE 12.5 Scenes to be created in order shown

You should now have several *Scene* tabs as shown in the image below (Figure 12.6). As you will see in a moment, SketchUp creates a nice smooth transition between *Scenes*.

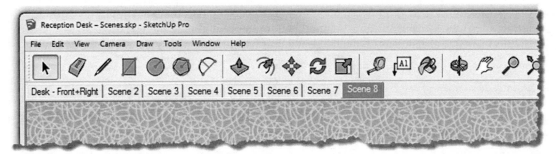

FIGURE 12.6 Scenes tabs added to the model

13. Click on each of the tabs, from left to right, to see how SketchUp transitions between tabs.

14. Now click on the tabs in a random order to see the transition results.

In both cases the transitions are nice and smooth. But, when we create the *Scenes* in a specific order we can better control the "path" for an animation. If required, we can use the *Scene Manager* to change the order of the scenes/tabs.

That is all you really need to know about saving scenes. Now let's talk about creating animations.

Creating an Animation

The first thing we will do is look at the settings related to animations.

15. Click on **View** ➔ **Animation** ➔ **Settings**.

This opens the *Model Info* dialog with the *Animation* category selected. Here you can specify how long it should take to get from one *Scene* to another.

This value can vary depending on the size of your model and the distance between the created *Scenes*. If the transition time is too short or the space

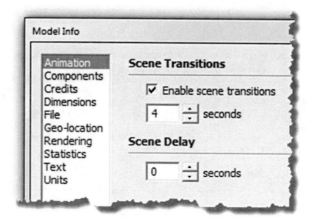

FIGURE 12.7 Model Info dialog

between the each *Scene* is too large, the animation can be way too fast. It takes practice to get everything just right so the animation looks good. This can be tricky because the transition time setting is for the entire animation. Sometimes you want to add additional *Scenes* to better control the path at corners or to stop and view an area – for example, looking from left to right without moving "your feet".

16. Change the *Scene Transition* time to **4 seconds**.

17. Change the *Scene Delay* to **0 seconds**.

> *TIP: This creates a smooth animation without a pause between each scene.*

18. **Close** the *Model Info* dialog.

19. Click on the first tab (named **Desk – Front+Right**).

SketchUp allows you to view the animation directly on the screen. We will try that first and then learn how to export the animation to a file which can be sent to a client or posted online for others to view.

20. To preview the animation, from the *View* menu, select **Animation → Play**.

21. Once you have seen the entire animation path, click the **Stop** button (Figure 12.8).

FIGURE 12.8 Animation controls

If you want to change anything, you either need to add new *Scenes* in the appropriate position or adjust an existing *Scene*. For the latter, you adjust the camera view as desired and then right-click on the *Scene* tab and click *Update*.

Now you will export the animation to an AVI file, which can be viewed by others using something like *Windows Media Player*. This way, the recipient does not need SketchUp installed, nor do they need to know how to use the software.

22. From the *File* menu, select **Export → Animation → Video**.

23. Click the *Options* button.

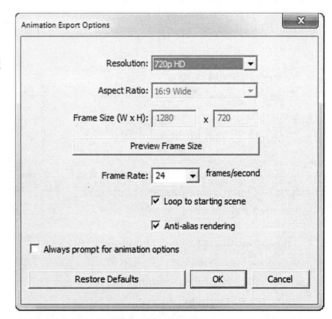

FIGURE 12.9 Animation export options

Here is where you control the quality of the animation (Figure 12.9). The higher the settings, the longer it takes to create the animation.

This author has worked on larger models/animations which have taken several hours for the export to complete. The **Width** and **Height** relate to screen resolution. You may have heard the team "HD" and "1080p" when talking about televisions. IF you set the height to 720 or 1080 you are making an animation equal to the quality of an HD movie. Setting the **Frame Rate** to 24 – 30 will also create a nicer animation. Finally, the **Anti-Alias** setting helps to smooth out angled lines.

For now, you will export the file with the default settings.

24. Click **OK** to close the *Options* dialog.

25. Browse to a location on your computer and click **Export**.

FYI: Notice the suggested file name for the MP4 is the same as your SketchUp model name.

You now should see the *Exporting Animation* progress bar (Figure 12.10). Depending on the complexity of your model, the length of the animation path, the resolution settings and the quality of your computer hardware, your animation time will vary. This example should take less than five minutes.

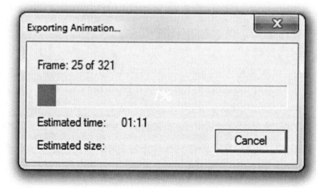

FIGURE 12.10 Exporting Animation progress bar

You can continue to work on other things on your computer as SketchUp is not utilizing all of your computer's resources to create the animation. You just cannot work on your SketchUp model.

26. Using *Windows Explorer*, browse to the location in which you saved your MP4 animation file.

27. Double-click on it to open the file and play it using your default video viewing software; this is typically *Microsoft Windows Media Player* (Figure 12.11).

The file should be about 19-20MB in size. This is getting a little large to email, but is no problem for FTP or a file sharing site such as DropBox.

Increasing the *Width* and *Height* values will make the initial view size larger on your screen.

FIGURE 12.11 Viewing animation in Windows Media Player

One more thing about animations you should be aware of: you can really only set up one animation path in your model. This creates a challenge when you want to develop multiple animations in a larger project. To do this, you actually need to create a copy of the entire file, delete the *Scenes* and create new ones. The real problem arises when you need to make changes to the model. Now you have multiple copies to manage.

Creating a Walkthrough Animation

Now that we have explored a simple animation example, let's take a look at creating a walkthrough in an office space. To do this, you will use a sample project provided by **LHB, Inc.** for their client **Hybrid Medical Animation**; both are located in Minneapolis Minnesota (Figure 12.12). The project was designed by Bruce Cornwall and the SketchUp model was created by Nick Vreeland. This project was an interior fit-out of an existing space. Using this model with give you a good feel for what you can create in SketchUp and how easy basic animations are to create.

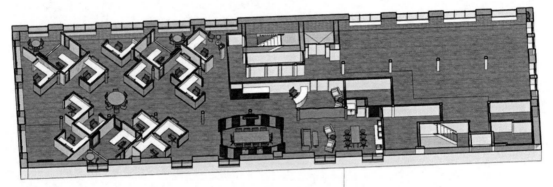

FIGURE 12.12 Sample project courtesy of LHB, Inc. and Hybird Medical Animation

28. Open the sample project, downloaded from the SDC Publications website, LHB **Project Sample - Hybrid Medical Anaimation.skp**.

You should now be in the *Interior View* scene tab.

29. **Right-click** on the **(Interior View)** scene tab and select **Scene Manager**.

Notice the option "Include in animation" is not checked (Figure 12.13). This tab will not be used in the animation. Both tabs are set this way in this sample project.

FIGURE 12.13 Scene manager

30. On the *Camera* toolbar, click the **Walk** tool.

31. While holding down the **Alt** key, drag the cursor around on the screen to walk through the model.

SketchUp, by default, will not walk through walls or large objects. It will also automatically go up and down stairs as it can go over smaller objects. However, when you have a dense model it can be hard to navigate. In these cases, you hold down the ALT key so collision detection is deactivated. This is also needed when you want to go through a wall or closed door.

32. When you are done exploring the space with the **Walk** tool, click the **Interior View** tab to restore that view.

Next you will create a series of *Scene* tabs to create an animation through the model.

33. Select **View → Animation → Add Scene**; or right click on the Interior View *Scene* tab and select **Add…**

This will be the first scene in the animation.

34. In the *Scenes* window, check the "**Include in animation**" option.

35. Use the **Walk** tool, plus **Alt**, to move forward - just before the round meeting table (Figure 12.15).

36. Create another **Scene** tab.

You will not rename each Scene tab as that would become too cumbersome. We will just know that the numbers indicate the order in which the animation will be created.

37. **Walk** forward, so you are directly over the round meeting table. And then create a new scene tab.

FIGURE 12.15 Creating scenes for animation through space

At this point in the animation, you want to left to the right. You will use the **Look** tool, which allows you to change where the camera is looking without moving it. Then you can create a new *Scene*, even though the camera location has not changes.

38. Select the **Look** tool, on the *Camera* toolbar.

39. Drag the cursor from right to left to look at the yellow wall (Figure 12.16).

40. Create another *Scene* tab.

41. Use the **Look** tool to look forward again, without moving, and save another *Scene* tab.

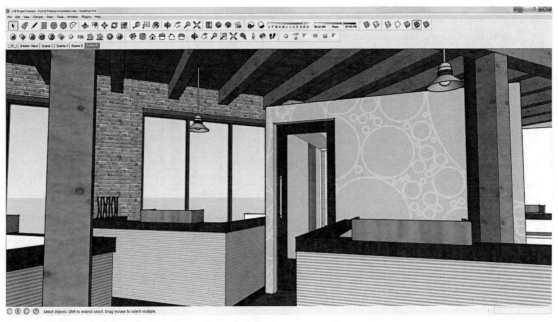

FIGURE 12.16 Using the Look tool to adjust the view

42. Continue to use a combination of **Walk** and **Look** to progress forward through the model. Turn around in the break room and work your way back to where you started (Figure 12.17).

43. Select **View → Animation → Settings**.

44. Change the *Scene Delay* to **0** so you have a smooth animation without any pauses.

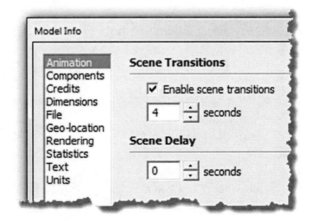

FIGURE 12.17 End the animation back at the start point—looking the opposite direction

45. To preview the animation, from the *View* menu, select **Animation → Play**.

46. Once you have seen the entire animation path, click the **Stop** button (Figure 12.8).

47. Select **File → Export → Animation → Video**.

48. **Browse** to a location to save you MP4 file.

49. Click **Options** and adjust the settings if you wish.

50. Click **Export**.

Some other things you may want to explore: you can add a section cut and save it to a scene. If one scene has the cut and another does not, SketchUp will animate the section cut. When transitioning between scenes. You can turn shadows on in an animation (setting must be saved with scene) but this can make interiors rather dark and the shadows sometimes have a flicker appearance, which can be distracting. This concludes our investigation of *Scenes* and *Animations*!

Sample project courtesy of LHB, Inc. and Hybird Medical Animation

Project Example: Custom Casework.

Image courtesy of LHB; www.LHBcorp.com

Chapter 12 Review Questions

The following questions may be assigned by your instructor as a way to assess your knowledge of this chapter. Your instructor has the answers to the review questions.

1. Each tab across the top of the drawing window represents a saved scene. (T/F)

2. When exporting an animation, you can adjust the overall quality. (T/F)

3. The Look tool allows you to change the view of the model without moving the camera. (T/F)

4. Animations can only be previewed once exported to a file. (T/F)

5. Scene tabs have nothing to do with the path an animation takes. (T/F)

6. You can only setup one animation path within SketchUp. (T/F)

7. A Scene tab may be excluded from an animation. (T/F)

8. Adjusting the Scene Delay setting to 0 will make a smooth animation. (T/F)

Chapter 13
Floor Plans

You will start out drawing the line work for the walls and openings. You do not need to create all the detail such as the doors and windows themselves, or the stairs and elevators. Rather you can start out by just sketching the walls which define these spaces. Text may be added if needed to remind you that a space is an elevator or stair.

The dimensioned drawing on the next page is what you will work from. The dashed line work at the rear (north) of the building are existing items intended to be removed. The two toilet rooms (northwest corner) are not accessible so they will be removed and replaced. Keep in mind, when renovating an existing building, the space below (and sometimes above) often needs to be accessible to route piping and wires in the ceiling. This can be tricky when that space is actively occupied. Additionally, in multi-story buildings it is often necessary to align the toilet rooms from floor to floor to minimize piping, drainage and clearance issues. None of this will be assumed to be a problem for our project.

The second means of egress is a fire escape on the back side of the building (Figure 13.1). This, along with the doors, will be removed and a new stair shaft will be added by the building owner in conjunction with this project to bring the building up to code. Notice the windows have been filled in. It would be possible to open a few of them back up on the sixth floor if needed, except the northeast corner at the service elevator location (see the plan on the next page).

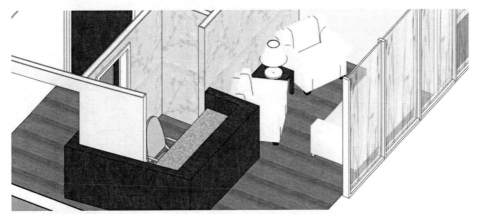

Project Example.   *Image courtesy of LHB; www.LHBcorp.com*

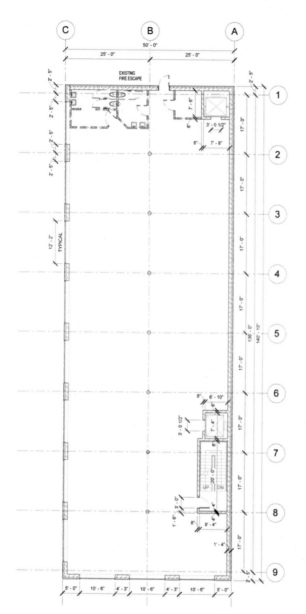

FIGURE 13.1 Existing conditions
See next two pages for enlarged floor plan views

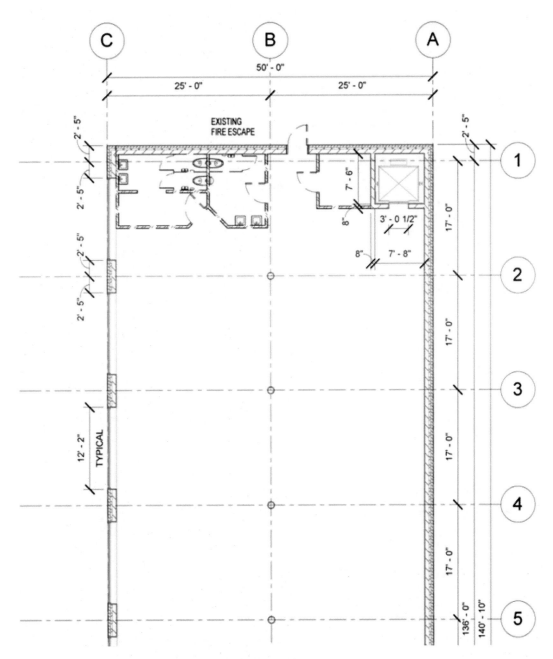

FIGURE 13.2 Existing conditions – enlarged plan - North

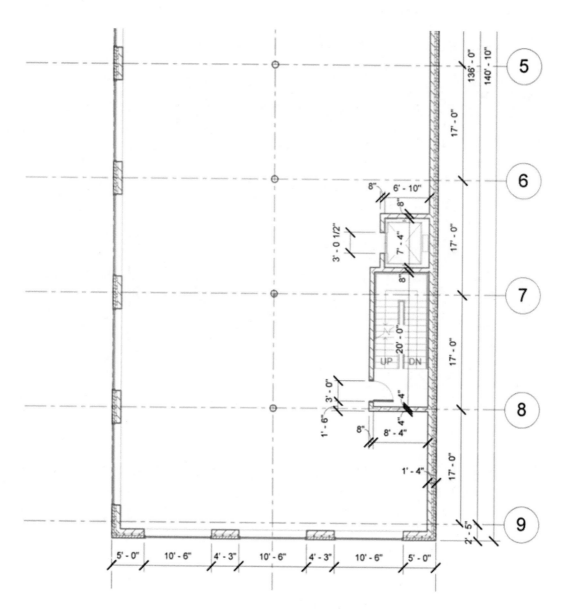

FIGURE 13.3 Existing conditions – enlarged plan - South

Now you will draw the existing plan, on the previous page, in SketchUp so you can print it out to scale on a plotter.

1. Using the steps previously covered in this book, create the **2D floor plan** in SketchUp:

   a. Follow the existing drawing on the previous pages (Figure 13.1 – 13.3).

   b. Set the **Camera** to **Parallel Projection** (see printing notes later).

   c. Switch to the top view (see printing note later in this chapter).

   d. All interior walls to remain are 8″ thick

   e. All exterior walls are 1′-4″ thick

   f. Only sketch the openings for the doors and windows.

   g. Do not draw the items to be demolished (shown dashed).

   h. Do not draw the lines for the stairs (i.e., handrails and steps).

   i. Do not draw the grids or dimensions.

   j. Add 12″ round columns.

When finished your SketchUp drawing should look similar to Figure 13.4. In the next chapter you will learn how to print this floor plan drawing to scale.

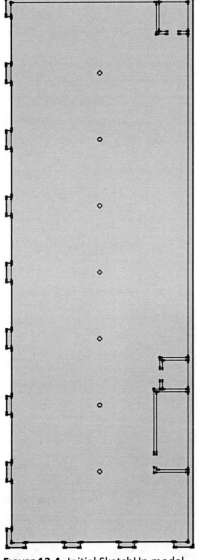

FIGURE 13.4 Initial SketchUp model

Notice the row of columns down the center of the building. These are a major component of the building's infrastructure and cannot be changed in anyway. They must be remembered and considered when working on design solutions.

In Figure 13.4, did you notice how the outside line of the exterior wall was omitted at each window opening? This is optional, and is used as a way to make all the window openings stand out more in this early stage of development.

All of the door openings are currently shown as simple openings in the walls. We have also left out anything we plan on demolishing: the existing toilet rooms and the door out the back of the building. These "demo" items will need to be drawn later in your CAD/BIM program to document for the contractor what, specifically, needs to be removed from the building. But for now that is not an issue we need to be concerned with.

For the simple plan we are working with here, it is not necessary to add any text. It is obvious that the two smaller rooms are elevators and the remaining set of walls defines the stair location, none of which can move because we are dealing with an existing building.

2. Save your SketchUp model as **Office Building Remodel.skp**.

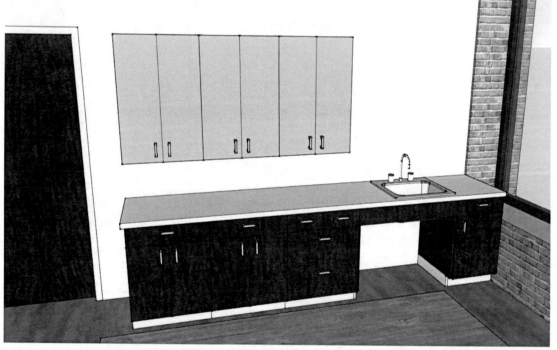

Project Example: Custom Casework.   *Image courtesy of LHB; www.LHBcorp.com*

Chapter 13 Review Questions

The following questions may be assigned by your instructor as a way to assess your knowledge of this chapter. Your instructor has the answers to the review questions.

1. This chapter covers drawing 2D floor plans from existing drawings. (T/F)

2. When sketching the round columns you entered the diameter. (T/F)

3. The camera setting needs to be Parallel, rather than Perspective for 2D drawings like this floor plan. (T/F)

4. Dashed lines (for walls, doors and windows) typically represents new construction. (T/F)

5. This floor plan matches the actual dimensions of the building. (T/F)

6. How thick are the exterior walls: _____ .

7. The portion of wall below the window has been ignored for clarity. (T/F)

8. Just get the view generally from above, no need to use the **Top** option in the *Views* toolbar.. (T/F)

Chapter 14
Printing Your SketchUp Model

The next thing you will want to do is print the floor plan. You have two paths you can take at this point. Both end up with the same result; that being a printed floor plan which is to scale.

One option is to print the floor plan on smaller paper, *maybe not even to scale.* This would allow you to do several quick sketch-overs by hand to start thinking about where spaces want to be and their adjacencies. An experienced designer can actually visually sketch pretty close to scale – picking up on known dimensions (e.g., the window width, stair width and depth, etc., are all known dimensions to the designer). Having a small printout is handy if you will be traveling or sitting somewhere and you have time to scribble a few ideas while you wait.

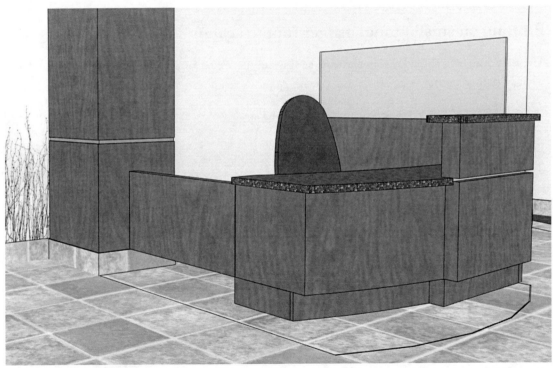

Project Example: Reception Desk.   *Image courtesy of LHB; www.LHBcorp.com*

This first option allows the designer to start thinking about the big picture aspects of the project. Quickly working through multiple scenarios allows the designer to rule in and out various ideas. Oftentimes the end result is somewhat serendipitous; that is, the final solution was derived by taking elements from several random iterations to come up with an ideal design solution. This final solution might never have otherwise been found if the designer spent too much time worrying about sketching to scale and trying to explore these early design scenarios in CAD/BIM/SketchUp.

Another option is to print the floor plan to scale and engage in a more formal sketch-over process at your desk. This route can be taken if you, or the client, have a fairly strong notion of what the layout will look like given the program and the space.

Next you will learn how to create both types of prints.

Printing on small format printer (not to scale)

The first thing you should do is adjust the settings in the *Print Setup* dialog. This tells SketchUp what printer or plotter you plan to use, plus the paper size and orientation.

1. Select **File → Print Setup** from the menu.

You are now in the *Print Setup* dialog (Figure 14.1). Your floor plan is tall and narrow so you will want a *Portrait* orientation. The default size is *Letter*, 8½" x 11". Most small format printers can also print on *Legal* size paper, which is 8½" x 14". Some printers will print on *Ledger*, which is 11" x 17".

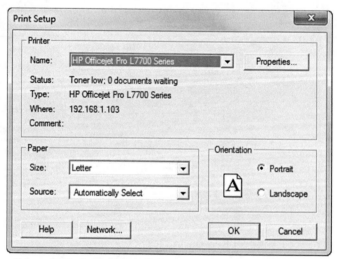

FIGURE 14.1 Print Setup dialog

Printer versus Plotter; what's the difference?

SketchUp can print to any printer or plotter installed on your computer.

A <u>Printer</u> is an output device that uses smaller paper (e.g., 8½"x11" or 11"x17"). A <u>Plotter</u> is an output device that uses larger paper; plotters typically have one or more rolls of paper ranging in size from 18" wide to 42" wide. A roll fed plotter has a built-in cutter that can – for example – cut paper from a 36" wide roll to make a 24"x36" sheet.

Color **printer** / copier

Plotter with multiple paper rolls

2. Select the following in the *Print Setup* dialog:

   a. *Printer:* (select any small format printer you have access to)

   b. *Size:* Letter

   c. *Source:* Automatically Select

   d. *Orientation:* Portrait

3. Click **OK** to accept the changes.

Before printing you will temporarily turn off the visibility of the faces as you do not need to see them on the print.

4. Select **View → Face Style → Hidden Line**.

The faces should now be hidden.

With the *Print Setup* options selected, these settings will be the defaults when you open the *Print* dialog box.

5. Select **File → Print** from the menu.

6. Make sure the following options are selected (Figure 14.2):

   a. *Printer:* already set based on *Print Setup*

   b. *Fit to Page:* Checked

   c. *Use Model extents:* Checked

   d. *Print Quality:* Standard

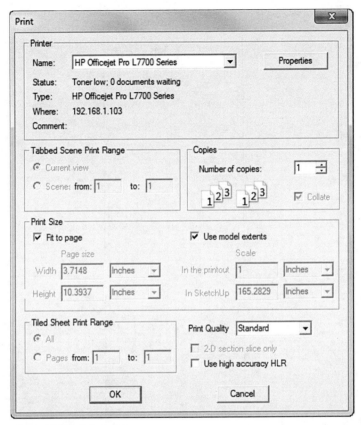

FIGURE 14.2 Print dialog

7. Click **OK** to print a copy of your floor plan.

At this point the *Print* dialog closes and your floor plan drawing is sent to the printer. It is a good idea to use the **Print Preview** option found in the *File* menu to visually verify what your print will look like before actually sending it to the printer. This saves time and paper! One common problem is with lines floating way out in space. When using the "Use model extents" option, this will cause the main model to be small as SketchUp is trying to include extraneous lines in the print.

Both your print and the *Print Preview* should look similar to the next image (Figure 14.3).

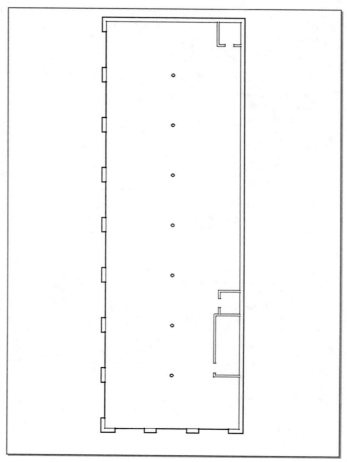

FIGURE 14.3 Print preview and "letter" sized paper

At this point you can use tracing paper to sketch over the floor plan printout or print a few copies and sketch directly on the printout.

You may want to write "not to scale" on the printout, or the letters "NTS", so you or others on the design team know the drawing is not to scale. It is possible that the printed drawing is really close to a standard scale and could create some major problems in terms of wasted time. This author has heard of contractor's getting their hands on NTS drawings and developing bids based them. Most contracts forbid scaling contract documents, but it still happens—so use caution!

Printing on large format plotters (to scale)

This section will show you how to print to a large format printer, also called a plotter. You will also learn how to print to scale and create a PDF file, both of which could also be applied to the previous steps on creating small format prints.

To actually print to a large format printer, or plotter, you need to have access to one. However, to work through these steps you will learn how to print to a PDF file. This process is identical to actually printing; the only difference is you select a PDF printer driver rather than an installed printer/plotter while in the *Print* dialog. Once you click **OK** to send the print, you are prompted for the file name and location. Very simple.

Most computers do not come with a PDF printer driver. If you have **Adobe Acrobat** (not *Adobe Reader*) installed you will have this PDF printer driver appear in your list of printers. Another option is to download a free driver from the internet. There are several options; however, Adobe does not provide a free option for this feature. One free option is PDF995 (PDF995.com). The free version displays small, self-promoting, company advertisements every time you print to PDF. There are several other options; just do an internet search to explore.

You will need access to a large format printer or a PDF printer driver in order to follow along with the next few steps.

8. Ensure you have set the following options:

   a. Camera → **Parallel Projection**

   b. Camera → Standard Views → **Top**

   c. View → Face Style → **Hidden Line**

In order to print to a specific scale you must have **Parallel Projection** and a standard view selected. Perspective views cannot be printed to scale.

9. Select **File → Print Setup**.

10. Make the following adjustments in the ***Print Setup*** dialog:

    a. *Printer*: select a large format printer or a PDF printer driver

    b. *Size*: 22"x34" (also called ANSI D)

    c. *Orientation*: Portrait

11. Click **OK**.

Next you will do a *Print Preview* so you can determine what scale will fit on the paper size you have selected.

12. Select **File → Print Preview**.

13. Make the following adjustments to the *Print Preview* dialog (Figure 14.4):

    a. *Fit to page*: **uncheck**

    b. *In the printout*: ¼ **(or .25)**

    c. *In SketchUp*: **12**

    d. *Print Quality*: **Large Format**

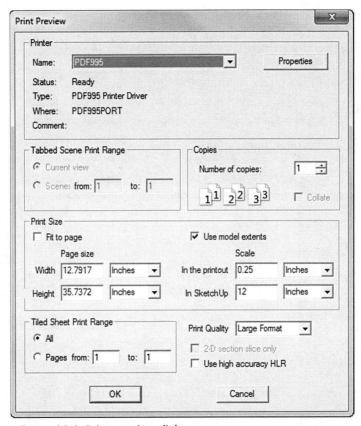

FIGURE 14.4 Print preview dialog

Notice, in Figure 14.4, that the page size for the printed drawing has updated when you specified a scale.

TIP: The page size does not update until you click into another field within the Print *dialog.*

Note that the *Height* is 35.7″, which is larger than your 22″x34″ paper you intend to print on. Therefore the ¼″ = 1′-0″ scale will not work.

FYI: If you printed with these settings, SketchUp *would tile the drawing onto multiple sheets of paper to print the entire drawing to the selected scale.*

Next you will adjust the settings to try another scale.

14. Change the **In the printout** option to **.125** in the *Print* dialog.

Notice the page size now fits on the desired paper size.

15. Click **OK** to see the *Print Preview*.

The preview image should look similar to Figure 14.5.

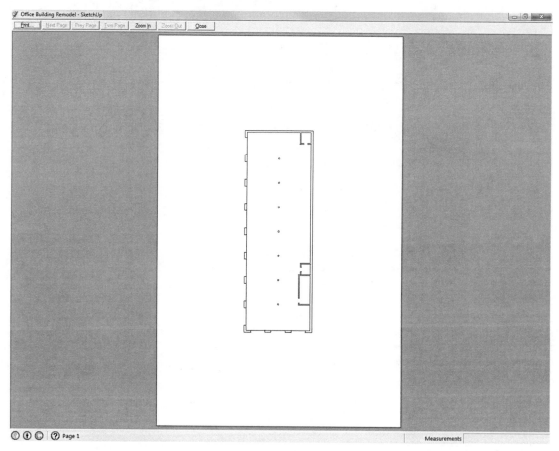

FIGURE 14.5 Print preview showing floor plan at 1/8″ = 1′-0″ on 22″x34″ paper

16. Click the **Print** button near the top of the application window to enter the *Print* dialog. Alternatively you can click *Close* and not print at this time.

17. Click **OK** to print the drawing.

You now have a print out or a PDF that is to scale. You can use your architectural scale on it to read existing building dimensions or sketch proposed items.

Exporting Image Files

Before moving on, we will mention another way in which you may output your model. You can export your current view to a raster image file. SketchUp supports many different file formats. This file can easily be emailed to a client or printed.

These are the steps to export an image:

- File → Export → 2D Graphic

- *Export Type:* Set this to the type of file you wish to create (jpg, png, etc.).

- Click the **Options** button and increase the pixels, as needed, to create a higher resolution image (Figure 14.6); the options vary slightly depending of file format.

- Specify the location and name for the new file.

FIGURE 14.6 Export image options dialog

- Click **Export**.

The exported file can be used in a report or presentation using Microsoft Word or Excel. It can be emailed or opened in an image editing program, such as Adobe Photoshop, to be enhanced or embellished.

Sample project courtesy of LHB, Inc. and Hybird Medical Animation

Chapter 14 Review Questions

The following questions may be assigned by your instructor as a way to assess your knowledge of this chapter. Your instructor has the answers to the review questions.

1. It is not possible to print drawings to scale from SketchUp. (T/F)

2. A plotter can only print up to 11"x17" size paper. (T/F)

3. You can export the current view on screen to a raster image file. (T/F)

4. Floor plan-type drawings need to be set to parallel projection. (T/F)

5. SketchUp does not have a *Print Setup* option. (T/F)

6. When entering the scale, for printing, you enter it in the *Print Preview* screen. (T/F)

7. SketchUp can tile a drawing image across multiple sheets of paper when printing. (T/F)

8. The Print Preview feature can save time and paper. (T/F)

Chapter 15
Creating the 3D Model

With the framework of the floor plan established you can quickly turn it into a 3D model using the tools and techniques previously covered.

1. Open your **Office Building Remodel.skp** file.

2. On the *View* toolbar, select the **Iso** icon to switch to a 3D view.

3. Click the ***Push/Pull*** tool on the toolbar.

Next you will pull upward the enclosed wall areas, making them 12'-0" tall.

4. Press and drag, upward, the corner shown in Figure 15.1.

    a. Release the mouse button once the wall is a few feet tall.

    b. The specific height is not important.

5. Type **12'** and then press **Enter**.

6. Repeat this for all enclosed wall areas (Figure 15.2).

7. Using ***Push/Pull***, drag the floor slab downward.

    a. Set the thickness to **5"**.

Now you are ready to work on the window openings. As you get more proficient with SketchUp, you will find much faster ways to do this. But for practice on the fundamentals, you will do everything manually and multiple times.

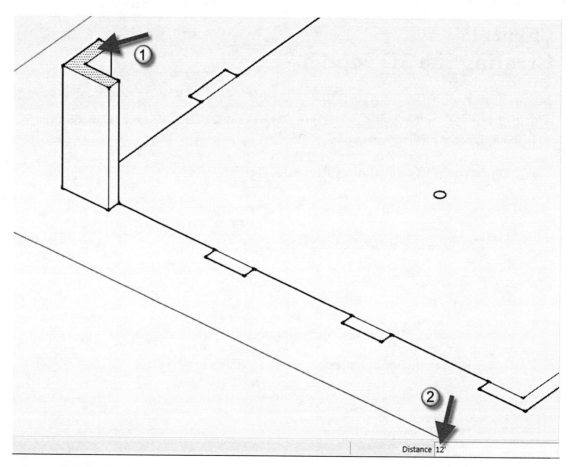

FIGURE 15.1 Raising the walls

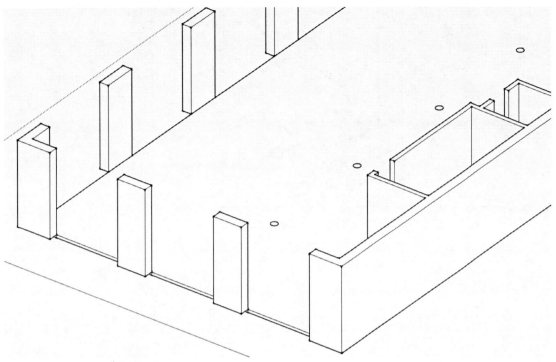

FIGURE 15.2 Raising the walls

8. Zoom in on one of the front windows.

9. Use the **Line** tool to draw a line from the floor straight up 3'-0".

   a. See picks 1 and 2 in Figure 15.3.

10. Draw a **Line** 10'-6" to the East.

    a. Step 3 in Figure 15.3.

    FYI: 10'-6" is the width of the opening; use the Tape Measure *tool if needed.*

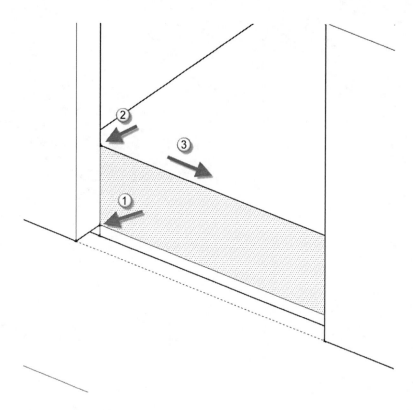

FIGURE 15.3 Adding window sill

You should now have a face which can be extruded out to create the portion of wall below the window. We will make this half the thickness of the main wall.

11. Use *Push/Pull* to create an 8″ thick wall below the window (Figure 15.4).

12. Repeat these steps for all exterior window openings.

As we continue to develop our preliminary plan, we will add a simple representation for the window and the portion of wall above.

The detail of the window can be developed later as the design becomes more refined.

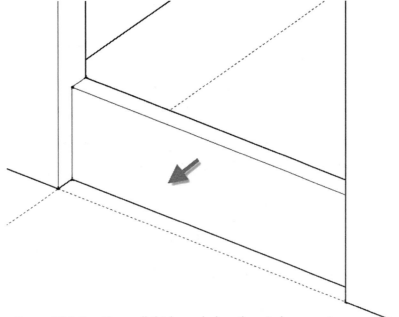

FIGURE 15.4 Creating wall thickness below the window opening

13. Draw two horizontal lines at each window opening (Figure 15.5):

    a.   One at **5′-0″** above the bottom of the window (#1).

    b.   The other aligned with the top of the wall (#2).

14. Use ***Push/Pull*** to create the **8″** wall above the window (Figure 15.6).

15. Repeat these steps for each window location.

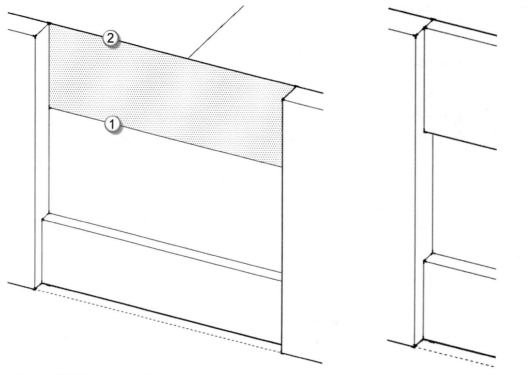

FIGURE 15.5 Creating wall above the window opening

FIGURE 15.6 Wall thickness

Once you get to the last window, the open edge of your model will be completely enclosed so SketchUp will create a surface across the top (Figure 15.7). You do not want to cap off your model at this point so you will delete this face.

16. Use the **Select** tool to select the face closing off the top of the model.

17. Press **Delete**.

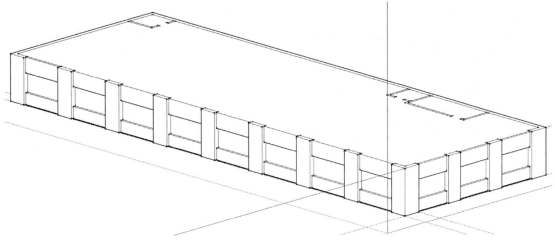

FIGURE 15.7 Windows added

The wall above the window needs a little work now that the "roof" is gone.

18. Use the ***Push/Pull*** tool to pull the wall back to the interior side of the wall.

   a. Do not pull the bottom part upward. Just select the back side of the wall and pull it back to align with the interior side of the main wall. While dragging the wall thickness back, move your cursor over the inside corner to snap to it.

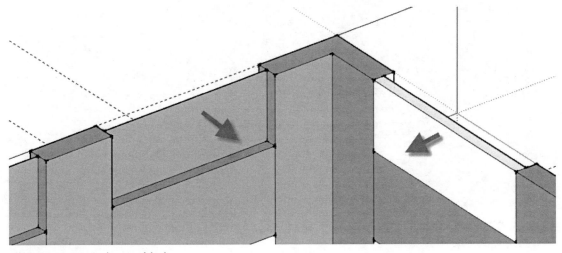

FIGURE 15.8 Windows added

Next you will add a few materials to round off this lesson.

19. Select **Windows → Materials**.

20. Add *Materials* to your model (Figure 15.9):

    a. Wood flooring

    b. Two colors on walls

    c. Sketchy material on top of walls

21. Turn **Shadows** on via the *View* menu.

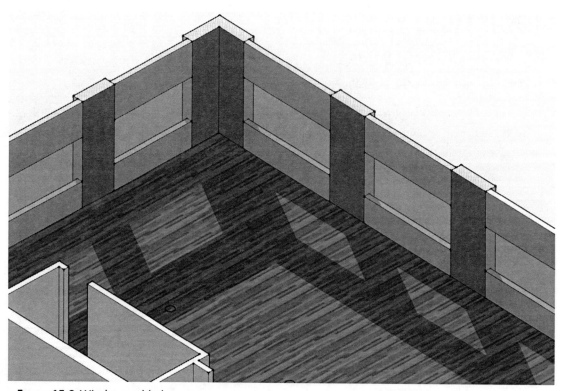

FIGURE 15.9 Windows added

22. Select **Camera** ➔ **Perspective**.

23. Select **Camera** ➔ **Position Camera**.

 a. Click where you want to be standing and then, whithout releasing the mouse button, drag towards the direction you want to be looking.

24. If you want to adjust ther view, drag the center wheel mouse button around and spin it to view the model from an interior perspective (Figure 15.10).

25. Select **View** ➔ **Animation** ➔ **Add Scene** to save this camera view.

26. **Save** your project.

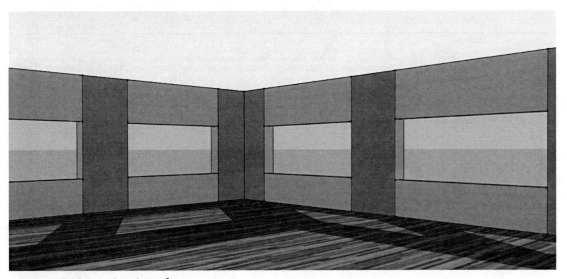

Figure 15.10 Interior view of space

As you develop the model, you can add more scenes (i.e. tabs) looking at different areas. As previously covered, you can also add scenses to develop a path for an animation.

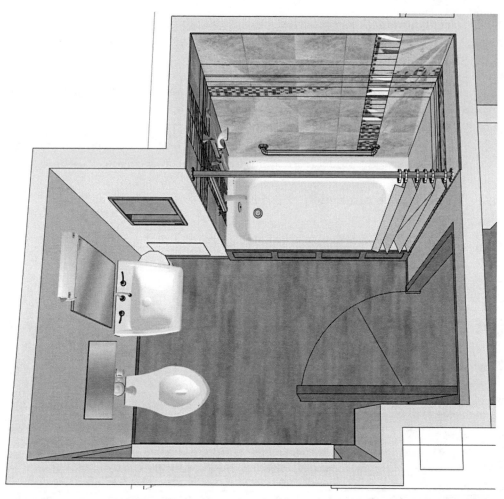

Project example: Walls Modeled.   *Image courtesy of LHB; www.LHBcorp.com*

Chapter 15 Review Questions

The following questions may be assigned by your instructor as a way to assess your knowledge of this chapter. Your instructor has the answers to the review questions.

1. You need to enclose an area for a face to appear. (T/F)

2. Your 2D plan can quickly be turned into 3D using Push/Pull. (T/F)

3. How tall are your windows: _____ .

4. You should use the Tape Measure too to verify you have drawn things accurately. (T/F)

5. Faces can be deleted at any time. (T/F)

6. What is the file size of your SketchUp model: _____kb.

Chapter 16
Lobby with Reception Desk

In this chapter you will create a lobby space and place some of the models previously created in this space.

1. Start a new SketchUp model.

2. Sketch the outline of the space as shown below (Figure 16.1).

    a. The dimensions shown are 18', 14', 3' and 22'.

    b. You do not need to add the dimensions, but you may want to use the Tape Measure Tool to double check your dimensions.

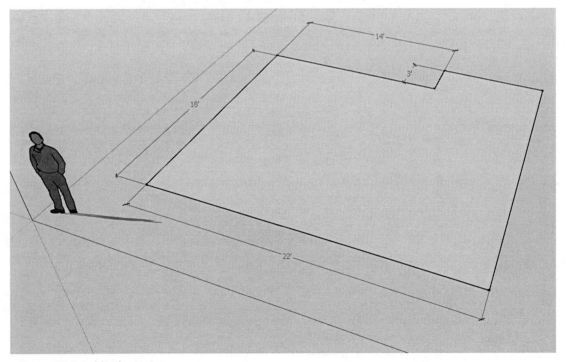

FIGURE 16.1 Lobby footprint

Next you will offset the perimeter 6" outward to define the wall thickness.

3. Select the **Offset** tool.

4. Offset the perimeter (Figure 16.2).

    a. Your first click needs to be on an edge (i.e. a line); wait to see the red dot on the line before clicking (as pointed out in the image below).

    b. Move your cursor perpendicular, away from the room.

    c. Click anywhere to finish the displacement.

    d. Type **6"** and then press **Enter** to adjust the offset.

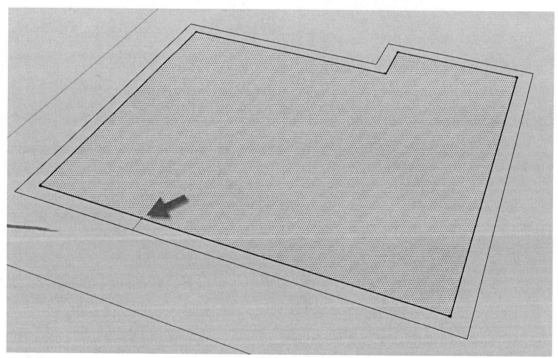

FIGURE 16.2 Offset perimeter to define wall thickness

5. Use **Push/Pull** to extrude the walls up **9'-0"** (Figure 16.3).

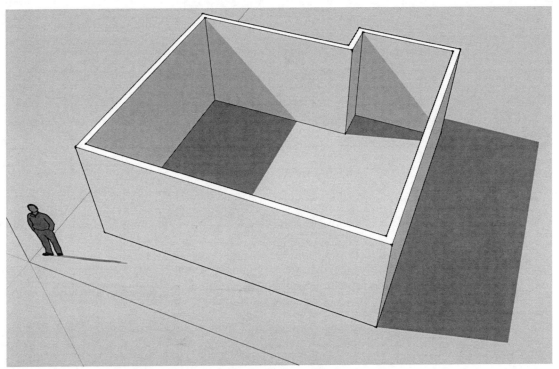

FIGURE 16.3 Use Push/Pull to create 9'-0" tall walls around the lobby space

Next you will create a window opening on the West wall.

6. Sketch the outline of the window opening (Figure 16.4)

   a. Start at the wall corner, drawing a line 3'-0" along the base of the wall

   b. Snap to the endpoint of the 3'-0" line and draw a vertical line 3'-0" up the face of the wall.

   c. Draw the 4'-0" x 10'-0" window opening.

   d. Delete the 3'-0" vertical line

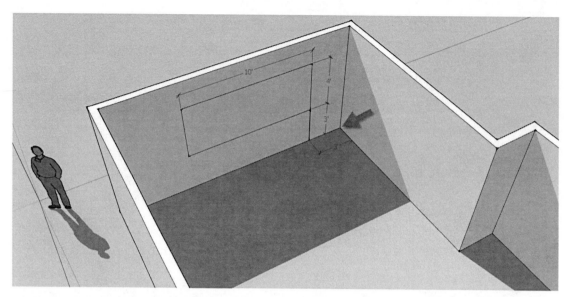

FIGURE 16.4 Sketching window opening

7. Use **Push/Pull** to create a hole in the wall for the window opening (Figure 16.5).

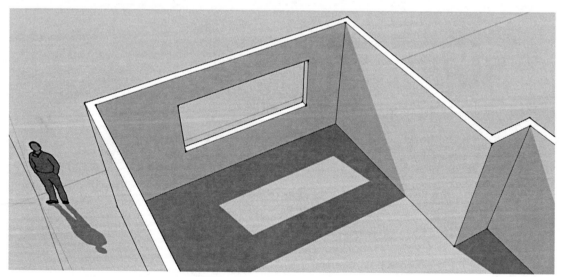

FIGURE 16.5 Window opening created using Push/Pull

Follow Me Tool

To quickly model the baseboard you will use the Follow Me tool. With this tool, you start by sketching the profile, or outline, of a shape you want to extrude. Next, you start the tool, click the profile to extrude, and then start moving your cursor along the path without clicking until you get to the end. This feature allows the profile to turn corners and follow curves all while creating clean transitions along the way.

8. Draw the profile of the 1" thick by 6" tall baseboard (Figure 16.6).

   a. Start somewhere along the north wall as shown

   b. Draw a line, on the floor, **1"** straight out from the wall

   c. Draw a line **6"** vertical

   d. Draw a **1"** line back to the wall

   e. Finally, draw a line **6"** down to the start point to complete the profile of the baseboard.

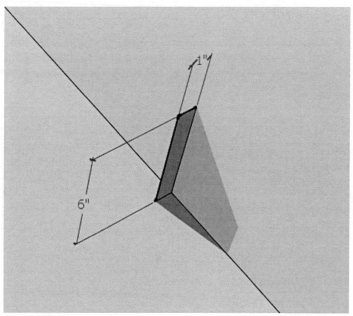

FIGURE 16.6 Profile of baseboard

Next you will activate the Follow Me tool to create the baseboard at the perimeter of the space. There are two ways to do this. The first is to select the Follow Me tool, click the face, within the profile of the baseboard, and then move your cursor along the path (even turning corners) and then click to finish the extrusion. The second option, and the one you will employ, is to select the floor first (whose edges define the perimeter of the space), and then start the Follow Me tool. Next you select the Face, within the profile of the baseboard, and you are done; the baseboard is instantly extruded alone the perimeter of the room.

9. Click the **Select** tool icon and then select the floor.

10. Select **Tools → Follow Me**

11. Selected the **Face** within the profile of the baseboard.

The baseboard should now be expanded around the perimeter of the room as shown in Figure 16.7.

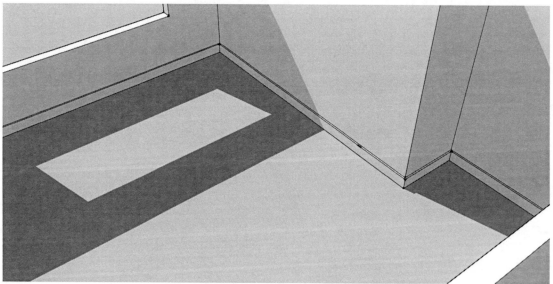

FIGURE 16.7 Baseboard added using the Follow Me tool

This was a very basic example, using a simple rectangle to define the baseboard profile. However, you can sketch any profile (Figure 16.8) and it will smoothly transition around corners as if it where miter cut Figure 16.9).

12. If you want, undo (using Ctrl+Z) and modify the baseboard profile to something more complex. Then, repeat the previous three steps.

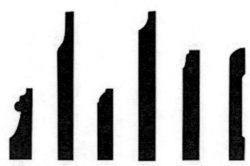

FIGURE 16.8 Baseboard profile examples

This tutorial will proceed with the baseboard shown in Figure 16.9.

Keep in mind that you can use this same technique to create cove moulding at the top of the wall.

Next you will use a similar technique to add trim around the window opening.

13. Zoom in on the lower right corner of the window opening; as viewed from inside the space.

> *TIP: Use the Pan tool (looks like a hand) and the Orbit tool to move around your model*

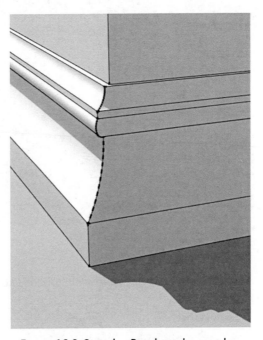

14. Sketch the profile of the window trim as shown; don't worry about the exact dimensions of the arc (Figure 16.10).

FIGURE 16.9 Complex Baseboard example

> *TIP: Be sure your lines are snapped to the red and green axes.*

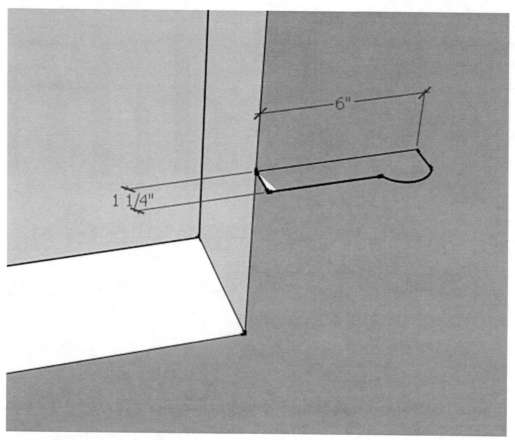

FIGURE 16.10 Window trim profile

Next you will use the Follow Me tool to extrude the trim profile around the window opening.

15. Select all the edges on the inside face of the window opening (Figure 16.11).

    a. First, click directly on one edge to select it

    b. Holding the Ctrl key, select the remaining edges

    c. Sketching the trim profile split the edge line, so be sure to select both edges

FIGURE 16.11 Select the edges of the window opening

16. Select the **Follow Me** tool and then select the face of the window trim.

Your trim is now extended around the perimeter of the window opening as shown in Figure 16.12. You can use this same technique to add a door opening into the room, with the same trim.

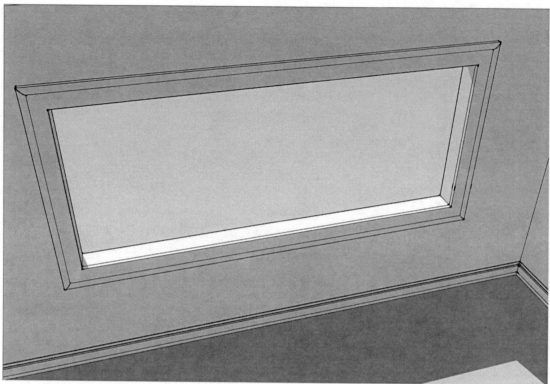

FIGURE 16.12 Complex trim profile extruded around window opening using Follow Me tool

Importing other SketchUp models

Now that you have the room fairly well defined, you will begin to layout some of the required elements of the space. We will use our reception desk and some of the furniture we previously created. SketchUp provides an **Import** tool, which allows you to insert another [entire] SketchUp model within the context of your current model. The result is a component which acts as a single entity (it is actually a component). This allows you to easily move it around without having to make sure you selected every line and face (which could be thousands). If needed, the component may be **exploded** so you can work with the base elements.

17. Select **File → Import** to initiate the import command.

18. Change the *Files of type* setting to **SketchUp (*.skp)** as shown in Figure 16.13.

Take a minute to notice the various types of files which can be imported into your SketchUp model. Note that some of these, such as the AutoCAD DWG format, is only supported by the professional version of SketchUp, not the free *Make* version.

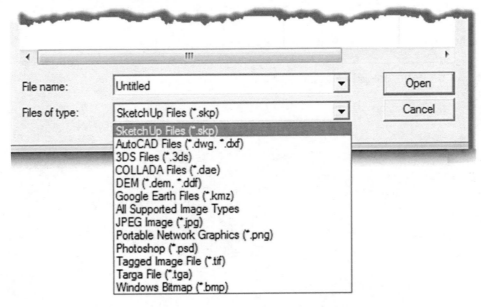

FIGURE 16.13 File types which can be imported into SketchUp

19. **Browse** to the location of your reception desk model previously created.

20. With the reception desk model selected, click **Open**.

21. Click to place the reception desk.

> **TIP:** *Make sure you see the **On Face** tooltip while your cursor is over the floor; this will ensure the reception desk ends up on the ground rather than in the air.*

If you still have a section plane defined in your reception desk model, the newly placed component will appear cut. Toggling off the *Section Cuts* in this model will affect the imported model.

22. Toggle **View → Section Cuts** if needed to see the entire reception desk.

23. **Move** the desk approximately as shown (Figure 16.14).

Notice how the entire reception desk model acts like a single component.

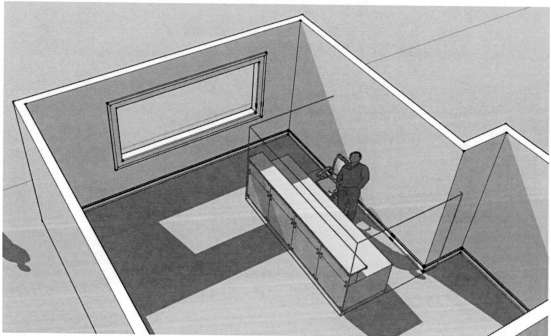

FIGURE 16.14 Reception desk imported into lobby drawing

We will leave the desk intact as a component to make it easier to move around when needed. If we need to make any changes to it, we can right-click on it and select **Edit Component**. Another right-click option for the desk is **Explode**; which will reduce the component down to its basic components of edges and faces.

24. Select the original SketchUp person in this model, near the origin; it is too strange to have two identical people in the model!

Add Materials

At this point we might want to begin adding materials to the space. First we will add a few predefined materials and then we will create a custom material. The custom material will be for the wall behind the reception desk; this will be a custom digital wallcovering by Sherwin-Williams.

25. Open the **Materials** window, if not already open.

26. Select **Tile** from the drop-down list.

27. Select **Tile Ceramic Natural** from the available options.

28. Click to **Paint** this material on the floor (Figure 16.15).

29. Switch the drop-down list to the Colors-Named material collection.

30. Select **0009_Linen** (near the top of the list).

31. **Paint** this material on all the walls within the space.

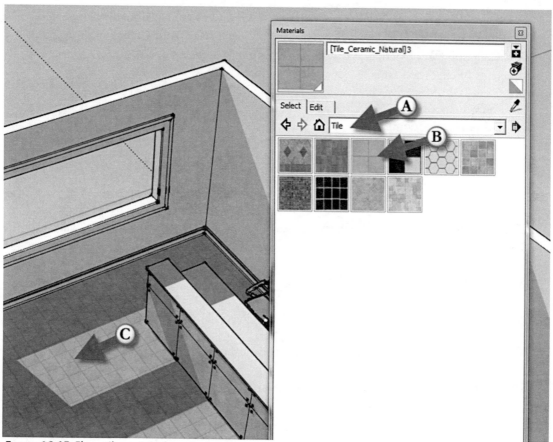

Figure 16.15 Floor tile material applied

Next you will add a material to the trim. This is a little trick given complex profiles we used for the trip. However, if we select all the faces for the trip we can then quickly paint them in larger chunks. The trick is selecting the faces without selecting adjacent faces (or behind). In this case, we can drag a select **crossing-window** (i.e. picking from right to left) in one of the corners (picks A and B in Figure 16.16). This will also select the adjacent wall and floor faces. To remove those faces, click directly on then while holding down the **Shift** key (picks 1-3 in Figure 16.16).

32. Select the baseboard in one corner as described in the previous paragraph.

33. From the Colors-Named material category, paint **0057_DarkKhaki** onto the selected baseboard.

34. Repeat these steps for the remaining baseboard and window trim (Figure 16.17).

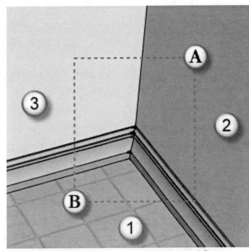

FIGURE 16.16 Selecting the baseboard faces

FIGURE 16.17 Painting a material on the trim

Your model now has a material on all surfaces within the model. Later you will add a ceiling surface and apply a material to it. For now we will leave it off so we can more easily access the model. However, it is possible to hide a face so it is not in your way; more on this later.

Next you will add the custom wall covering to the wall behind the reception desk. The first thing you need to do is download the image.

35. **Browse** to the following website:
    http://surrounddecor.com/custom-digital-wallcovering/swirls.html

36. **Click** the preview image below with just the pattern (pointed out in Figure 16.18).

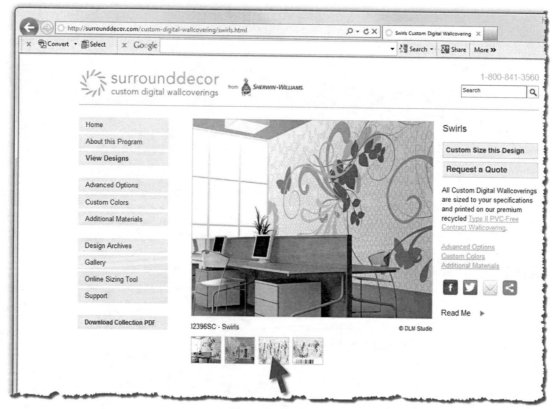

FIGURE 16.18 Downloading image from wallcovering website

37. Right-click on the image and select **Save picture as...** from the pop-up menu (Figure 16.19).

38. **Save** the image file to your computer's hard drive.

FIGURE 16.19 Save the image to your computer

Next you need to create a custom SketchUp material so you can paint this image on a face.

39. In the *Materials Window*, click to **Create material** icon in the upper right.

40. Name the material: **Swirl**.

41. Check the option to **Use texture image**.

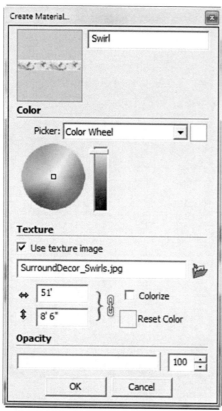

FIGURE 16.20 Creating new material

42. **Browse** to the image file you just saved.

43. Adjust the <u>height</u> to **8'-6"** – this is the height of the wall, above the baseboard.

44. Click **OK** to create the new material.

45. Use the **Paint** tool, to apply the new Swirl material to the wall directly behind the reception desk.

The image is now applied to the wall, but it needs to be repositioned (Figure 16.21). Depending on the quality of the image you downloaded, the image may not look too good. In this case you may need to contact the manufacturer and see if they can send you a high resolution image file.

FIGURE 16.21 Custom material applied to wall

46. Click the **Select** tool, and then **right-click** on the face with the swirl image.

47. Select **Texture → Position** from the pop-up menu.

You now see the extents of the image (8'-6"x51'-0") and then the image tiled in each direction (Figure 16.22). At this point you can drag the image around to get the desired portion of the image to appear on the wall.

FIGURE 16.22 Adjusting the position of the image

48. Move the image as desired; be sure the top of the image aligns with the top of the wall.

49. Apply the same material to the return wall, and then use the *Position* feature to align the texture at the corner; this may take a little trial and error Figure 16.23).

You can use this technique in many situations in your SketchUp model to enhance the model and better convey your design intent.

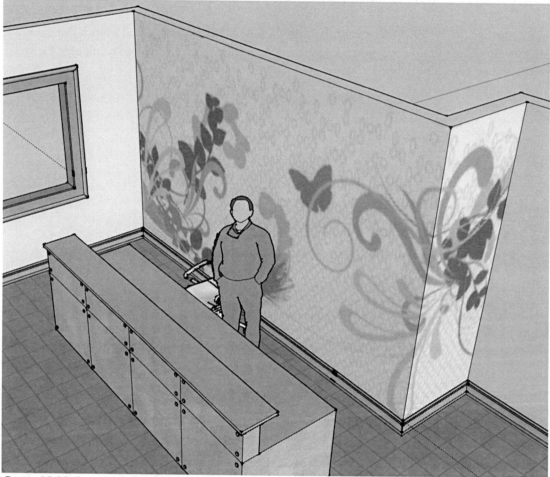

FIGURE 16.23 Custom wallcovering wrapping corner

Before we bring this chapter to a close, we will add a few more elements to the space. We could add the furniture we created earlier in the book, or you can go online and download some more complex items.

50. Select **File → 3D Warehouse → Get Models…**

51. Type **potted plants** in the search window and press **Enter** (Figure 16.24).

You now have several options of plants which can be loaded directly into your model. The size and quality can vary greatly as this content is largely from "end users" of the software like this author and the readers of this book.

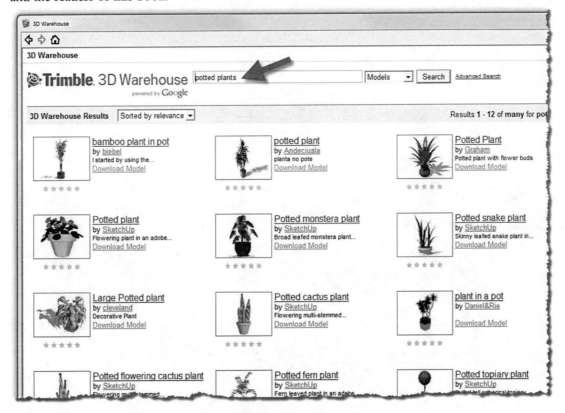

FIGURE 16.24 3D Warehouse model browser

52. Click one of the images, and then click **Download Model**.

53. Click **Yes** to the prompt: **Load directly into your SketchUp model?**

If the file take too long to load (i.e. it is way too complex) or you don't like what you see, you can click *Undo*.

54. Position the plant on the floor, approximately as shown in Figure 16.25.

55. Add lobby furniture; in 3D Warehouse, search for: **Steelcase Turnstone – Social Hub**. Place off to one side, right-click and Explode. Now move portions into your space—arrange items as desired. Delete any unused elements.

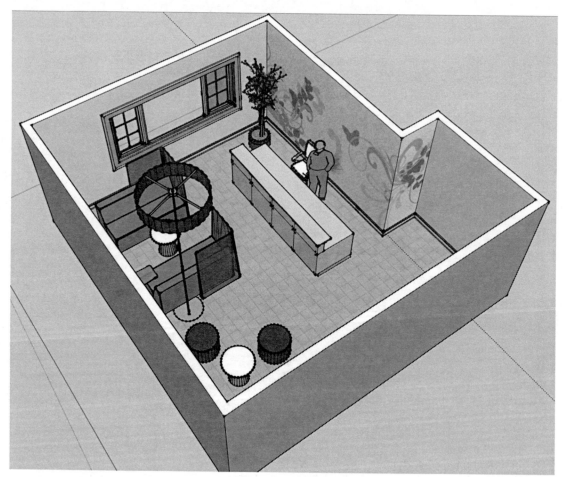

FIGURE 16.25 Plant, lobby furniture and window added to model

The last step in this chapter will be to add a window into the window opening. Again, you will search 3D Warehouse for options.

56. Search 3D Warehouse for **Picture Window**.
    a. The component used in this example was an option one the first page of results (called Windows Picture plus mullioned sides 3 by 2.
    b. The component may need to be placed off to the side of your model, **Rotate** and then **Move** into the window opening.
    c. Once in the window opening, use the **Scale** tool (if needed) to size the window to match the opening. Hold **Shift** when dragging the grips.

57. Save your model as **Reception Desk in Lobby.skp**.

Sample project courtesy of LHB, Inc. and Hybird Medical Animation

Chapter 16 Review Questions

The following questions may be assigned by your instructor as a way to assess your knowledge of this chapter. Your instructor has the answers to the review questions.

1. Use the **Explode** tool to reduce a *Component* to it individual elements (i.e. editable edges and faces). (T/F)

2. Holding the *Shift* key while clicking on a face will de-select it.. (T/F)

3. Use the *Import* tool to bring in other SketchUp models. (T/F)

4. SketchUp Pro can also import AutoCAD DWG files. (T/F)

5. A section cut within an import SketchUp model does not do anything. (T/F)

6. What tool allows you to quickly extrude a shape (or profile) around a room:

   _____ _____ .

7. When placing the reception desk, make sure you see the _____ _____ tooltip while your cursor is over the floor.

8. An imported SKP file becomes a *Group*. (T/F)

Chapter 17
Photo-Realistic Rendering in SketchUp

This chapter will take a look at one way in which the capabilities of SketchUp may be expanded by 3rd party developers; namely photo-realistic rendering. SketchUp is widely known for its sketchy, non-realistic presentation capabilities. However, it is possible, with the aid of an extension, to create some amazingly realistic images, both exterior and interior, within SketchUp. Doing everything, or at least as much as possible, in one program can save a lot of time.

There are so many add-in options SketchUp has an **Extension Warehouse**, similar to *3D Warehouse*. These extensions are developed by software companies who want to offer specific functionality not already built-into SketchUp. Take a few minutes to explore the Extension Warehouse by clicking on the same named icon on the toolbar. Notice there are nearly 50 (at the time of this writing) rendering related extensions.

This tutorial will look at a specific extension called V-**Ray for SketchUp**. This tutorial is not meant to, in any way, encourage the reader to buy this product; this author has no connection with the makers of this software. It is possible to work through this chapter using the fully functional 14-day demo available from: www.chaosgroup.com.

Hybrid Medical Animation, Minneapolis MN   *Modeled by Nick Vreeland—LHB, Inc.*

Your model should be roughly 6MB at this point. The first thing you will do is save a copy in case anything goes wrong you can quickly start over at this point.

1. Open your lobby model, and select **File → Save-As**.

2. Save your model as **Reception Desk in Lobby – Rendering.skp**.

Next you create two scene tabs so you can quickly move between a top-down view and an interior view.

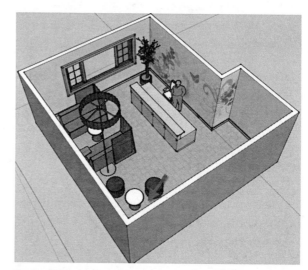

3. Using **Pan**, **Zoom** and **Orbit**, adjust the view to look similar to Figure 17.1.

4. Select **View → Animation → Add Scene**.

5. If you get a *Scenes and Styles* warning, just click **Create Scene**; leaving "Do nothing" selected.

FIGURE 17.1 Saving a scene

Next you will position a camera within the space and then save another scene.

6. Select **Camera → Position Camera**.

Where you click will be where you want to be standing within the model

7. Click a point on the floor, in the approximate area pointed out in Figure 17.1.

8. Select **View → Toolbars… → Camera**.

9. If you are not looking towards the potted plant, use the **Look Around** tool, from the *Camera* toolbar, so you are.

The *Look Around* tool allows you to just look around the model without moving your "feet". Because the space is rather small, you will want to adjust the **Field of View** of the camera to see more of the room; this is analogous to using a wide angle lens. This adjustment is done while the regular *Zoom* tool is active.

10. Select the **Zoom** icon.

Notice the status bar indicates the *Shift* key can be employed to adjust the field of view.

11. Hold down the **Shift** key and drag your cursor downward. Notice the degrees listed in the lower right.

12. Get the degrees near 60 and then release the mouse button.

13. Type **60** and then press **Enter**.

14. Use the **Look Around** tool to adjust the view to look similar to Figure 17.2.

FIGURE 17.2 Positioning camera within space and adjusting the field of view

15. Select **View** → **Animation** → **Add Scene...**

You now have two scene tabs which can be used to quickly navigate the model. Next you need to add a ceiling. When creating photo-realistic renderings, the model needs to be completely enclosed so you are not getting any extra light in the space.

16. Click the **Scene 1** tab to switch to the isometric view.

17. At the top of a wall, on the interior side, use the **Line** to sketch a line on top of another line.

This will create a face across the top of the space; i.e. your ceiling.

18. Click the **Scene 2** tab to switch back to the interior view.

19. Open the **Materials** window, if not already open.

20. Paint **010_Snow**, from the *Colors-Named* category, onto the ceiling face.

Most ceilings are white to help reflect light in the space. When using indirect lighting this is a must. The ceiling will still appear somewhat gray, rather than bright white. This is due to the way SketchUp treats different adjacent planes to better convey depth.

Next you will adjust the isometric view so the ceiling is hidden.

21. Switch back to the **Scene 1** tab.

22. Click the **Select** icon, and then click the ceiling to select it.

23. **Right-click** on the ceiling and select **Hide** from the pop-up menu.

If you were to click the *Scene 1* tab again, the ceiling would re-appear. If you want this change to stick, you need to update the Scene associated with the tab. You will do the next.

24. **Right-click** on the **Scene 1** tab and select **Update**.

25. Switch between the two tabs to see the ceiling visibility change.

Setting up Daylighting

Daylighting is an important design aspect. Next we will look at some basic settings for the sun. We will assume the true north is straight up. It is possible to georeference your model for the project location on earth.

26. Make sure the proper toolbar is on: **View → Toolbars → Shadows**

This toolbar allows you to turn shadows on and off as well as set the month and time of day.

27. Click to turn on shadows and set the month and day as shown above.

With shadows turned on, the light from the sun is also visible for interior views. You should see light on the reception desk and floor as shown in Figure 17.3.

When the photo-realistic rendering is being generated, the light from the sun is actually bounced around the room. This real-life simulation helps to create a more realistic effect.

FIGURE 17.3 Shadows turned on

V-Ray for SketchUp

The remaining steps in this chapter require **V-Ray for SketchUp** by the *Chaos Group*. If you do not have access to this extension, you can install the 14-day demo and use that to complete this chapter. The demo version may be found here: http://www.chaosgroup.com/en/2/vrayforsketchup.html

There are additional rendering extensions which can produce similar results. You may wish to explore those if you wish. However, the options and settings for each are extremely different. So, to follow along, you will need to be using V-Ray for SketchUp.

If you don't have access to this software, you can skip to the next chapter. You will not be missing anything, as the final result in a photo-realistic rendering is a raster image file. The preparation often involves tweaking the model for best results. However, for this basic introduction, we have already made all the changes to the model itself.

The makers of this extension have provided the following description of their offering:

> SketchUp® users in all fields depend on V-Ray® as a quick, easy and cost-efficient way to render their most cutting-edge images. Developed by the Chaos Group, V-Ray for SketchUp works with Trimble's SketchUp, one of the most popular 3D modeling tools available today.

When V-Ray for SketchUp is installed, you will have two additional toolbars:

- VfS: Lights
- VfS: Main Toolbar

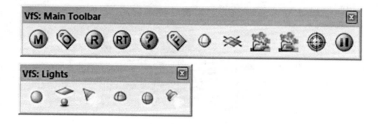

These toolbars may be docked at the top of the screen with the other toolbars. You have access to nearly all of the V-Ray tools via these toolbars. These tools can also be accessed from the Plugins pull-down menu. A few additional settings are accessed when right-clicking on certain elements within the SketchUp model.

To show how easy and powerful the basics are for this extension, we will simply apply a set of Presets and the hit render!

27. Click the V-Ray **Options** icon.

In the **Options Editor** you could get completely lost in the settings and dialog boxes. This level of complexity is what allows designers to develop some amazing images to help convey their designs. However, to get really decent results we do not have to be an expert. One can simply apply one of several sets of presets.

28. Click the **Load Defaults** icon to ensure all settings are at their defaults as shown in Figure 17.3..

29. Set the *Presets* to **Interior** and **04_High_ Quality_Interior**

30. Click the **green checkmark** icon to apply the presets.

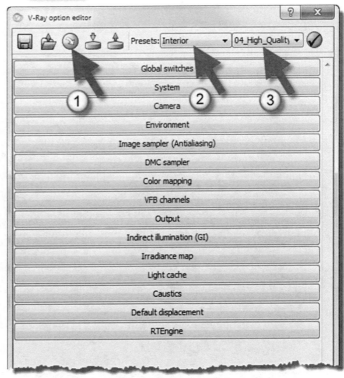

FIGURE 17.4 V-Ray option editor

You are now ready to render your first image. You do not have to close the V-Ray option editor. Adjustments are applied immediately. As you can see, there is no "ok" or "apply" buttons. You now ready to render the scene.

31. With *Scene 2*, the interior view, current—Click the **Start Render** icon.

Within a few seconds you should have a nicely rendered image as shown in Figure 17.5. **If you are using the demo version**, the resolution is limited and the "V-Ray" watermark is applied (as shown in this first example below). The image is a little dark, but overall is not too bad! It will be much better by the end of this chapter.

FIGURE 17.5 V-Ray initial rendering

Clicking the **Save Image** icon in the rendered window allows you to save this rendering to a file.

Now that we see how easy the basics are, we will explore some of the V-Ray settings and features to make the image even more compelling.

Here is a list of some of the things we will do to enhance the rendering:

- General lighting enhancements
- Add ambient lighting to the scene
- Add an artificial lights; at floor lamp and reception desk
- Make the privacy screens transparent
- Add a texture to the sofas
- Make the tile more dynamic
- Add a slight reflection to the wallcovering

With each adjustment made, we will re-render the scene. With the size of our model, we can create a full rendering with rather high settings to see the results. Each rendering will only take a few minutes. The rendering engine has the ability to use all the cores in a multi-core processor (CPU). This means the rendering process utilizes all the CPU computing power possible.

On larger projects, the final rendering process can take hours to complete. In this case the designer will use draft settings (e.g. low resolution output) to get a good idea of the final product; sacrificing quality for speed. Once everything generally looks good, you can then increase the quality settings for the final rendering. The settings needed will vary depending on how the final image will be used. If the image will just be used in a PowerPoint-type presentation, then the resolution does not need to be as high as would be needed for print, or—in a more extreme situation—a large banner or billboard application.

Another option on larger projects is the **RT Render** engine. In addition to using your computers CPU and RAM, the resources on your graphics card (i.e. GPU and RAM) will also be engaged. This allows the designer to see really high quality draft images in just a few minutes.

General Lighting Enhancements

The first thing we want to do is brighten the scene because it is too dark. It is possible that we will have to revisit these settings. Once the proper reflectances (which helps light bounce around the space) are added to the materials and artificial lights are employed, the scene may end up too bright.

There are two simple settings to quickly increase the ambient light levels in the scene; **Shutter Speed** and **Film Speed (ISO)**. If you are an amateur photographer, or better, you will recognize these terms relate to settings on a camera. Seeing as our goal is to create a <u>Photo</u>-realistic image, it makes sense that we would have real-world type camera settings available.

32. In the *V-Ray option editor*, expand the camera settings panel by clicking on the **Camera** bar.

    FYI: *You do not have to close the render window, but you can if it is in the way.*

33. Adjust the *Shutter speed* to **30** (Figure 17.6).

We will not change need to change the film speed setting in this case. Also, notice when you hover your cursor over the settings a descriptive tooltip appears.

34. Click the **Start Render** icon again to re-render the scene.

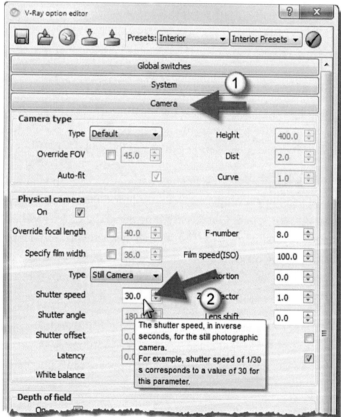

FIGURE 17.6 Adjusting the Camera settings

As you can see the image is much brighter now (Figure 17.7).

FIGURE 17.7 Image brighter after adjusting shutter speed to 30

Another general lighting adjustment we will make is to soften the edges of the sunlight.

35. Click the **Camera** bar to hide those settings.

36. Click the **Environment** bar, and then click the "**M**" button next to *GI (skylight)*

37. Set the **Sun: Size** to **8.0** (Figure 17.8) and then click **OK**.

38. Run the rendering again (via the **Start Rendering** icon).

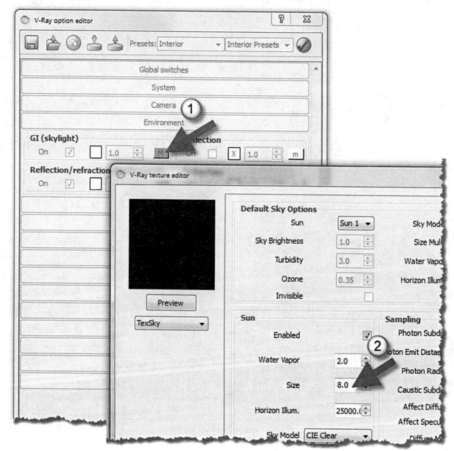

FIGURE 17.8 Adjustment to soften the edge of the sunlight hotspots

The edge of the sunlight is now softened (Figure 17.9).

Ambient Occlusion

Adding ambient lighting, aka, **ambient shadows**, helps to ground objects in space and give a better sense of depth to the space. It can also help to better define edges of objects.

39. In the *V-Ray option editor*, click on the **Indirect illumination (GI)** bar.

40. Check the box to turn on **Ambient occlusion** (Figure 17.9).

41. Rerun the rendering.

As you can see, there is now much more definition in the view (Figure 17.11). With just a few simple settings the scene is already a lot better looking.

FIGURE 17.9 Softened edges of sunlight

Global Illumination (GI):
V-Ray for SketchUp uses GI to simulate the sunlight. Even when you don't have artificial lights added to the model, you can still get a decent rendered image. This is not unlike what you could do with a camera in the middle of a bright day. Many rendering programs create vary dark interior images unless you add artificial lights and take several additional steps to setup daylighting.

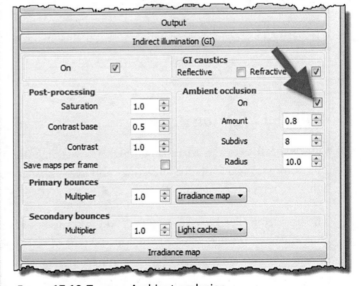

FIGURE 17.10 Turn on Ambient occlusion

FIGURE 17.11 Scene rendered with ambient occlusion turned on

Artificial Lighting

Our next task is to add artificial lighting. Artificial lighting is manmade light fixtures and bulbs. These are essential for night renderings and help enhance realism in daytime scenes seeing as some lights are typically on during the day.

The first artificial light source you will add is in the floor lamp on the left. You will edit the component so the light becomes an integral part of it. If your SketchUp project had multiple instances of this component, they would have the light source.

42. **Right-click** on the floor lamp and select **Edit Component**.

Everything in the model is grayed out except the component. You will now use the *V-Ray for SketchUp* tool called **Sphere Light** to place the light source. Remember, because you saved the scene, you can zoom and pan around to better see various parts of the component. When finished, you can simply click the **Scene 2** tab to get back to your rendering view.

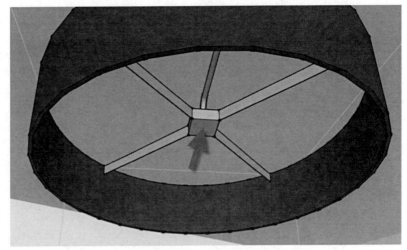

FIGURE 17.12 Adjusting the view to modify the floor lamp

43. Using the scroll-wheel on your mouse, adjust the view of the floor lamp as shown in Figure 17.12.

> *TIP: Holding the Shift key while pressing the center scroll-wheel allows you to pan the view. This is a quick way to utilize both Pan and Orbit.*

V-Ray offers a few different ways to add artificial lighting to your model. You will try two of the most basic options: *Sphere Light* and *Rectangular Light*. You can also create **spot lights**, with controls to focus the light, and **IES lights**. An IES light is based on the mathematical definition of a real-world light figure. Most lighting manufactures have these IES files available for download from their websites. This is what lighting analysis programs, such as **AGI32**, use to calculate point-by-point illumination values required by some codes and performance standards such as LEED®.

FIGURE 17.13 Sphere light tool

44. Select the **Sphere Light** tool (Figure 17.13).

45. Pick two diagonal points as shown (Figure 17.14).

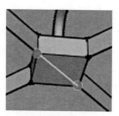

FIGURE 17.14 Pick two diagonal points to create the Sphere light tool

The light source has now been added to the component. It may be a little large, or not; depending on the design. We will leave it as-is, but if it needed to be scaled down, you would just use SketchUp's *Scale* tool.

46. **Right-click** on the sphere light and select and select **V-Ray for SketchUp → Edit Light** (Figure 17.15).

You now see the properties for the selected light source (Figure 17.16). We will not make any changes at this time, but note the **Intensity** can be changed to brighten or dim if needed. You can also make the source **Invisible** if you did not what to see the "bulb".

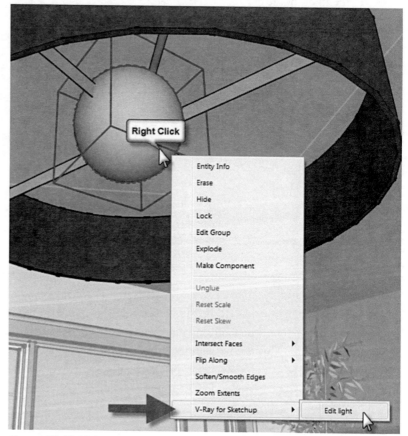

FIGURE 17.15 Light source added to floor lamp component

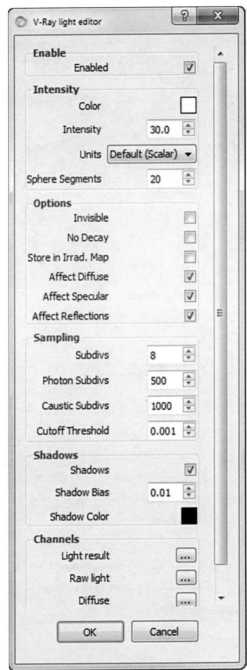

FIGURE 17.16 Light source properties

47. Click **OK** to close the *V-Ray light editor* window (Figure 17.16).

48. Right-click anywhere and select **Close Component**.

49. **Render** the scene to see how the added light source affects the final image.

    Notice in the image below (Figure 17.17, you can see a hotspot on the ceiling form the light source. Also, the top of the sofa is also brighter. Depending on your view, you may be able to see the light source.

FIGURE 17.17 Light source properties

Next you will add a rectangular light source to suggest an under-counter light at the reception desk.

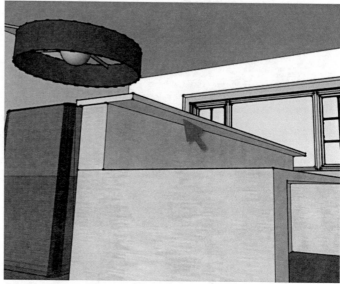

FIGURE 17.18 Adjust view as shown

50. Adjust the view so you see the underside of the transaction counter, as shown in Figure 17.18

51. Select the **Rectangle Light** tool.

52. Pick two diagonal points under the transaction counter to create a rectangular light source which covers the entire area pointed out in Figure 17.18.

Initially this light source is only emitting light in one direction, which is likely in the wrong direction (i.e. up into the countertop). Next you will adjust the light properties to make it double sided and decrease the light intensity a little.

53. Right-click on the rectangular light and select **V-Ray for SketchUp → Edit Light**.

54. Change the following (Figure 17.19):

    a. Intensity: **15.0**

    b. Double sided: **Checked**

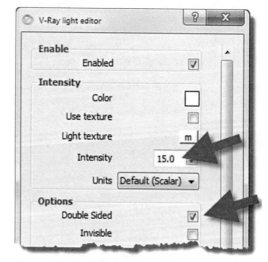

FIGURE 17.19 Adjust light settings

55. Click **OK** to close the light editor window.

56. Reset the view by clicking the **Scene 2** tab.

   TIP: You can rendering any view on the screen if you want to see the results from another angle.

57. **Render** the view.

Notice the countertop now has light on it (Figure 17.20).

FIGURE 17.20
Light on countertop

Making Privacy Panels Transparent

In the SketchUp model we can see through the privacy panels in the waiting area. However, we cannot see through them in the renderings we have created thus far. We need to make an adjustment to the material in the **V-Ray material editor**. This special material editor gives us many extended and advanced options not available in SketchUp alone.

Before we can edit the material, we need to find out which material is being used.

58. Right-click on the privacy panel and select **Edit Component** (#1; Figure 17.21).

59. Right-click again, and select **Edit Component**; this is a nested component.

60. Select the SketchUp paint bucket tool, and then click the **eyedropper** from the *Materials* window (#2; Figure 17.21)

61. Click on the privacy panel face.

62. Note the material listed (#3; Figure 17.21).

FIGURE 17.21 Privacy panel: edit component and discover material used

Now that we now which material to edit, we can open the V-Ray material editor.

63. Click the **V-Ray Material Editor** icon.

You are now in the V-Ray material editor. On the left you can see all the materials currently defined in the SketchUp model. Stretch the window wider if you cannot read the material names. The right side shows the properties for a selected material.

64. Select **<auto> 10** in the material list.

65. Right-click and select **Rename Material**.

66. Rename to **Privacy Panel**.

67. Click the **M** icon next to *Color*, and note the path for the texture_1.jpg file. Should be something like:
    C:\Users*username*\AppData\Local\Temp\skp20131014182837\models\untitled\
    texture_1.jpg

68. Close *Color* settings window, and then uncheck **Use Color texture from transparency** (#2; Figure 17.22).

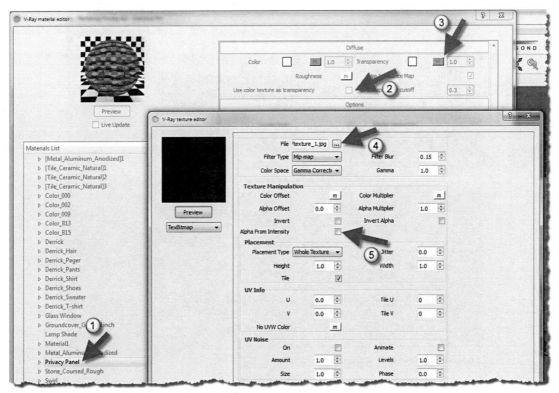

FIGURE 17.22 Making privacy panel material transparent

69. Click the **M** icon next to *Transparency* (#3; Figure 17.22).

70. Click the "**...**" icon next to *File*; browse to the same JPG file found in step 67.

71. Uncheck **Alpha From Intensity** (#5; Figure 17.22).

72. Click **OK** to close the window and then close the *V-Ray material editor* window.

73. **Render** the scene.

The privacy panels are now somewhat transparent (Figure 17.23). Essentially, we used the same texture which defines the material to punch holes in the material. The default settings would just make the entire material transparent. That would be great for glass, or maybe stained glass, but not they fabric.

FIGURE 17.23 Privacy panel material is now transparent

Add a Texture to the Sofas

Next we will turn our attention to sofas. We want the texture to look more realistic. There are a number of ways to do this.

74. Use the steps previously covered to discover the material used for the sofas; you will learn it is **<auto>**.

75. Open the *V-Ray material editor* and rename **<auto>** to **Sofa Material**.

76. Make the following edits to **Sofa Material** (Figure 17.24):

    a. Right-click on the material and select **Create Layer → Reflection** from the pop-up menu.

    b. Under *Reflection*, edit the **Glossiness** section (steps #2 and #3)
        i. Highlight: **0.8**
        ii. Reflect: **0.65**
        iii. Subdivs: **36**

    c. Under *Diffuse*, click the **M** icon next to *Color* (#4)
        i. Set the drop-down to **TexNoise** (#5; Figure 17.25)
        ii. Set **Color A** to Red: **8**, Green: **139**, Blue: **186** (#6)
        iii. Set **Size** to **0.25** (#7)
        iv. Select **Inflected Perlin** (#8)
        v. Click **OK**

    d. Check **Displacement** and then select the **M** icon (#9; Figure 17.26)
        i. Set the drop-down to **TexNoise** (#10)
        ii. Set the **Size** to **25** (#11)
        iii. Click **OK**

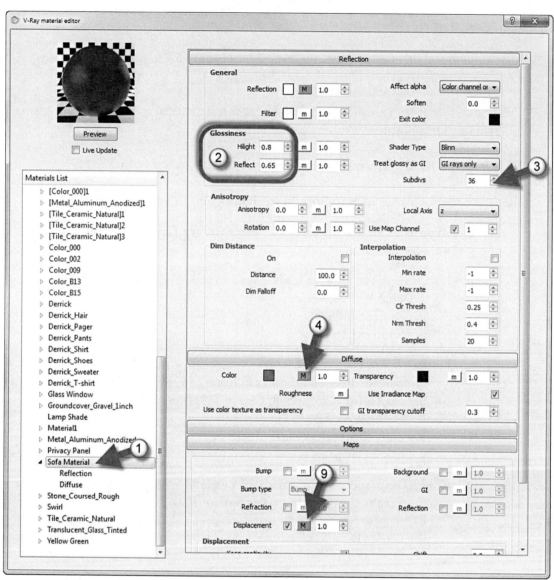

FIGURE 17.24 Adjusting sofa material to look like leather

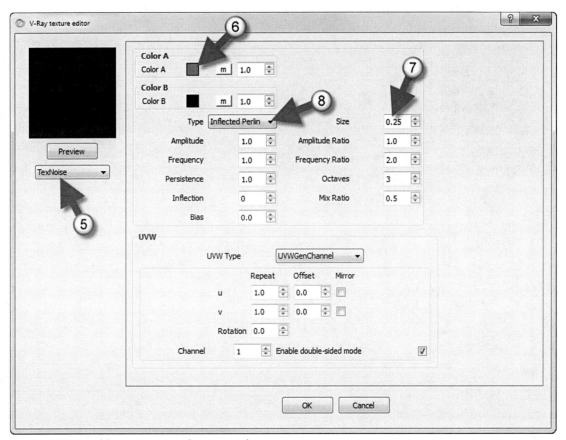

FIGURE 17.25 Adding texture to the material

Adding noise to the material will help the surface look less smooth. Adjusting the Type changes how the pattern is applied to the surface. Changing the size controls how large, or small, the patter is. If you were adding this as a stucco surface, the size might be larger.

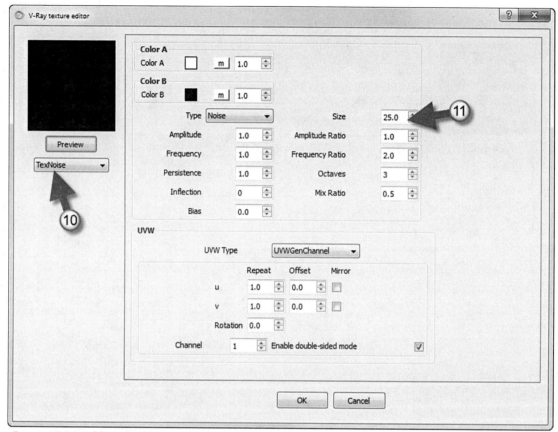

FIGURE 17.26 Adding displacement to make the surface more irregular

The *Displacement* feature can make a surface look three-dimensional. Here we are doing it with a built-in "noise" modifier to add a bit of a wavy look to the sofa (so it does not look perfectly flat, as it was modeled). This does not actually change the modeled geometry, just how it appears in the rendering. We can also use this feature to make masonry or stone walls look very realistic. In this case we might use a grayscale version of the original texture to define the high and low parts of the material.

77. **Render** the scene to see how the material comes out.

Notice how the material has a texture to it, a somewhat irregular surface and is more shinny looking (Figure 17.27). The color has shifted a little darker based on our settings.

FIGURE 17.27 Adding displacement to make the surface more irregular

Modifying the Floor Tile and Wallcovering

Next you will make a few adjustments to make the floor tile and wallcovering look more realistic. Before making any changes, run another rendering and look at the tile and wallcovering closely. Maybe even click the save icon in the rendered window to compare with the results of the changes you are about to make.

78. Use the **eyedropper** tool, on the *Materials* window, to verify which material you added to the floor.

79. *Rename* this material to **Lobby Floor Tile**

You will use the displacement feature to emphasis the irregular surface of the tile.

80. Make the following adjustments to the *Lobby Floor Tile* material:

    a. Right-click on the material name and select **Add Layer → Reflective**.

    b. Under *Reflection*, edit the **Glossiness** section
        i. Highlight: **0.8**
        ii. Reflect: **0.65**
        iii. Subdivs: **36**

    c. Check **Displacement** and then select the **M** icon
        i. Set the drop-down to **TexBitmap**
        ii. **Browse** for the primary raster image used for this material. It should be in a location similar to this: C:/Users/*username*/AppData/Local/Temp/ChaosGroupTextureCa che/[Tile_Ceramic_Natural]_extractedTex.jpg
        iii. **Check** the **Invert** option.
        iv. Click **OK**

    d. Set the value next to displacement to **12**.

81. **Render** the scene to see how the tile appearance has changed.

The floor tile now has more definition and highlights (Figure 17.29).

82. Using the steps previously covered; add a **Reflection** layer to the **Swirl** material. Set the multiplier value next to *Color* to **0.25** (Figure 17.28).

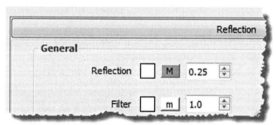

FIGURE 17.28 Reflection settings

83. **Render** the scene.

FIGURE 17.29 Final rendering using V-Ray for SketchUp extension

Your final image is now rendered (Figure 17.29). You are encouraged to save the file and then **save-as** a copy and make some adjustments to the various settings. The book cover image has the reveals in the front of the desk modeled (as the lines do not appear in the rendering) and additional reflectances added. If you want to learn more about V-Ray for SketchUp be sure to read through the user guide and check out the Chaos Group's website for additional learning options. You will also want to check out sketchucation.com.

Sample project courtesy of LHB, Inc. and Hybird Medical Animation

Chapter 17 Review Questions

The following questions may be assigned by your instructor as a way to assess your knowledge of this chapter. Your instructor has the answers to the review questions.

1. SketchUp can create photo-realistic renderings without any extensions. (T/F)

2. **V-Ray for SketchUp** is a free add-in for SketchUp. (T/F)

3. Some setting adjustments can have a significant impact on total rendering time. (T/F)

4. The rendering extension used in this chapter is the only one on the market for SketchUp. (T/F)

5. The quickest way to get started with **V-Ray for SketchUp** is to use the provided presets. (T/F)

6. GI stands for _____ _____ .

7. Adjusting the _____ and _____ have a major impact on the brightness of the rendered scene.

8. **V-Ray for SketchUp** uses SketchUp's sun settings. (T/F)

Chapter 18
Introduction to Layout

The professional version of *SketchUp* comes with a tool, called *LayOut*, which allows the *SketchUp* model to be composed on sheets for printing. These printed sheets can be for presentations or construction documents. When SketchUp first came out, in the year 2000, it was generally understood that this new tool would not replace the traditional CAD (i.e. AutoCAD, Microstation, etc.) software, but rather supplement it. Of course, time changes most everything; since then, not only has SketchUp ownership changed hands twice, they developed *LayOut*. Now, there are plenty of designers, architects and interior designers who don't even own a traditional CAD program. It should be pointed out, however, that *SketchUp* and *LayOut* are still not appropriate, in this author's opinion, for large complex projects such as a new hospital.

In this chapter, which requires SketchUp Professional, you will create one 11"x17" sheet with presentation views of your lobby model. Another sheet will be created to compose 2D-type construction drawings.

The *LayOut* portion of SketchUp Professional is actually a separate program. This program has its own desktop icon (as shown to the right). The model is sent to *LayOut* from within *SketchUp*.

LayOut 2013

Getting Started with LayOut and the User Interface

The first thing we will do is load your lobby model into *LayOut* and then review the *LayOut* user interface.

1. Open your lobby model in SketchUp. You can use the model from the previous rendering chapter, or the chapter prior to that if you did not do the rendering chapter.

2. Select **File → Send to LayOut**.

Once you click **Send to LayOut**, the *LayOut* application open (Figure 18-1). Here, you are presented with a templates dialog. This allows you to select paper size, if you want gridlines or a titleblock.

3.   On the left, expand **Paper** and select **Plain Paper**.

4.   On the right, in the preview area, select **Tabloid Landscape**.

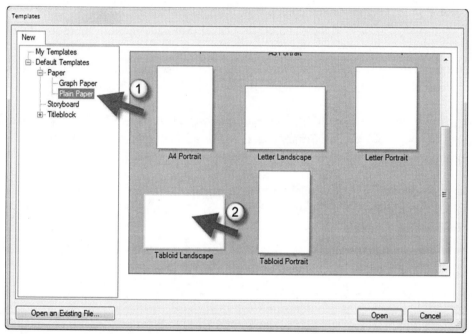

FIGURE 18.1 Templates dialog for LayOut

Tabloid is 11"x17" paper. This is a common size for smaller drawing sets and presentations. The titleblock section has larger sheet sizes which also include a titleblock (i.e. sheet border) for project information such as client name, project location, date, etc.

5.   Click **Open** to accept the selected template.

6. Select **View → Restore default workspace** to ensure your screen is similar to the image below.

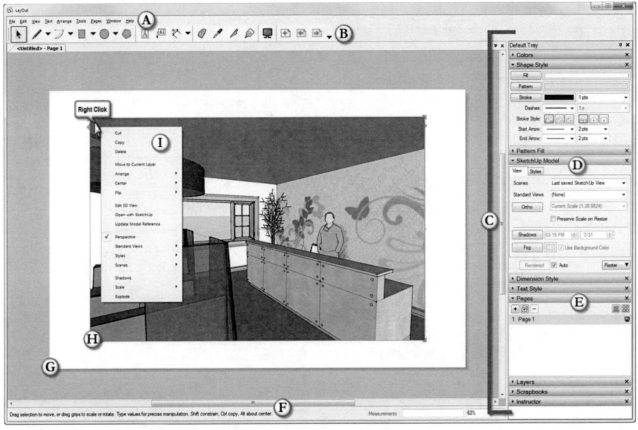

FIGURE 18.2 LayOut user interface overview

The following lettered items correspond to the letters in Figure 18.2 above.

A. Pull-down menus
B. Toolbars
C. Tool tray
D. SketchUp related settings
   (for current view)
E. Add, remove, rearrange pages

F. Status bar
G. Paper (11"x17" in this example)
H. View on paper
   (of referenced SketchUp model)
I. Right-click options
   (for view on sheet)

Let's elaborate on these items.

The **Menus** are **Toolbars** are functionally the same as in *SketchUp*. You can customize the tools, but there is only one toolbar. Clicking on a black down-arrow next to an icon will reveal additional, related, tools (see example to right).

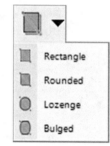

The **Tool Tray** area is similar to the tool windows, from the *Window* menu, which can be opened in *SketchUp*. However, in *LayOut*, these windows are docked to the side of the application window. Clicking the tool title bar will either expand or collapse that tool window. Clicking the "X" in the title bar will close the window. To make is visible again, turn it back on via the *Window* menu. Much of this information is contextual, in that you have to select something to see the values light up and become editable.

Within the *Tool Tray*, one important section is the **SketchUp Model** window. When a view is selected on a sheet, you can use this area to adjust the view using *SketchUp*-based settings. Some of these setting options are pulled directly from your *SketchUp* model. For example, all the saved Scene tabs are available in the *Scenes* drop-down list. Selecting one of these will update your view on the page to match the saved scene in the *SketchUp* model.

The **Pages** window, also within the Tool Tray area, allows you to quickly add, remove and rearrange pages. All pages in a *LayOut* project are the same size, so clicking the "+" icon to create a new sheet, will quickly create another sheet of the predetermined size.

The **Status Bar** provides insightful prophets for active tools or selected items. You also have the Value Control Box as in SketchUp. Finally, there is a drop-down to control the zoom factor of the page. Keep in mind the page represents a real-world size piece of paper (11"x17" in this example). Thus, the design is actually scaled down to fit on the paper. More on setting the drawing scale for printing later.

Item G, in the user interface image, is highlighting the rather obvious fact that you can discern the edge of the paper. Similarly, **item H** is identifying a view on a page (you can have multiple views on a page).

Finally, when **right-clicking** on a view you are presented with several options. Some of these options are also found in the *Tool Tray* to the right. Notice the option to select a scale; this allows each view to be set to a specific scale. When printed to scale, each view can be measured using an architectural scale.

LayOut Document Setup and Preferences

Here we will quickly look at a few settings "under the hood" in LayOut.

7. Select **File → Document Setup**

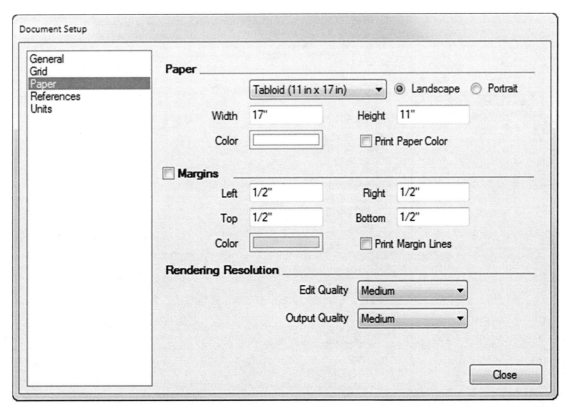

FIGURE 18.3 LayOut document setup dialog

Many of the initial settings here are derived from the template you choose and apply to the entire *LayOut* project you are working in. These settings can be changed at any time. Notice the **paper size** can be changed. Of course, changing from a larger sheet to a smaller sheet may require views to be move, re-scaled or deleted to fit on the smaller sheet. Another important setting here is **Units**. If you are working in the United States you will want this set to fractional—Inches. Finally, notice *LayOut* remembers where the original *SketchUp* model is; under **References**. This means you *LayOut* project will remain to-to-date with any changes made in your *SketchUp* model. Once a *LayOut* project is started, be careful not to move, or rename the SketchUp file name, or folders.

8. Click **Close** to exit this dialog box and the select **Edit Preferences**.

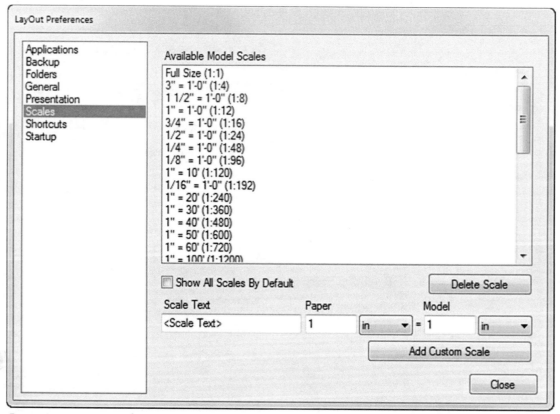

FIGURE 18.4 LayOut Preferences dialog

In the **Preferences** dialog there are several settings and options which relate to the program as a whole; not just the current document/project. Notice you can delete or add **Scales** here. Take a moment to explore a few of the other options without making changes.

9. Select **Close**.

Working with Viewports

Next you will resize the view, copy it and change the scene in the copied view.

10. With the *Selection* tool active, click on the view to select it.

11. While holding down the **Shift** key, drag the lower right corner of the view up, and to the left (to make it smaller); approximately as shown in Figure 18-5).

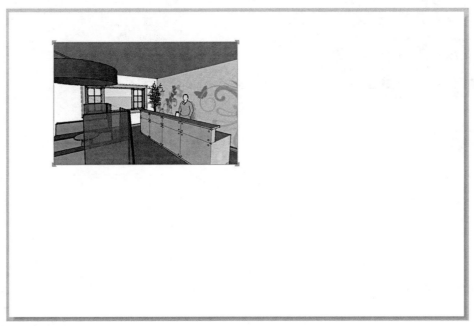

FIGURE 18.5 View adjustment on paper

12. Click and drag the view up and to the left a little more to reposition the view on the paper.

13. While holding the **Ctrl** key, drag a copy of the view to the right.

 TIP: While dragging the view, you will notice a dashed alignment line to make sure the views line up.

14. Adjust the two views, via Shift + corner drag, so they are the same size (Figure 18-6).

FIGURE 18.6 Duplicate and adjust view adjustment on paper

15. Select the image on the right.

16. In the *Tool Tray*, expand the *SketchUp* model panel (if not already) and then set *Scene* to **Scene 1** via the drop-down list (Figure 18.6).

Notice the two scenes you created, in *SketchUp*, are listed here.

Your page now has two views of your *SketchUp* model. Both of these views are tied back to the same *SketchUp* model. If the original *SketchUp* file is modified, the views on this sheet will also update.

Updates are down whenever the *LayOut* document is opened and by Selecting the **Update** button in the **Document Setup →** **References** area.

In addition to selecting predefined views from the *SketchUp* model, you can develop custom vantage points of the *SketchUp* model within *LayOut*. You will try that next.

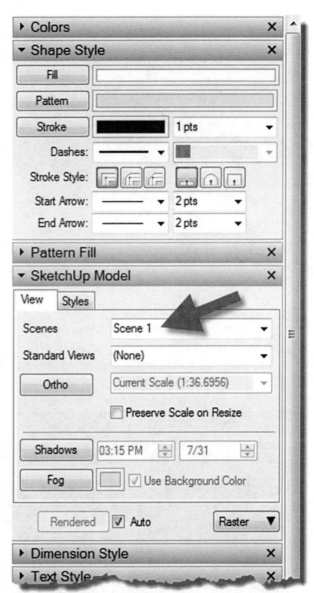

FIGURE 18.7 Switching scene setting for selected view

FIGURE 18.8 Two scene-based views of the SketchUp model in LayOut

17. Using **Ctrl + Drag**, copy original view, down, towards the bottom of the page. Initially it will not fit, which is ok.

18. Drag the corners, <u>without</u> using Shift, to fill the page as shown in Figure 18.8.

FIGURE 18.9 Third view added to the page

At the moment, the new third view is just a modified view of the initial view. Next you will activate the view and use the *SketchUp*-like **Orbit** and **Pan** tools to adjust the view.

19. Double-click within the view to activate it.

20. Adjust the view to look similar to the view shown below (Figure 18.10). Drag the center wheel button to **_Orbit_** and hold *Shift* for **_Pan._**

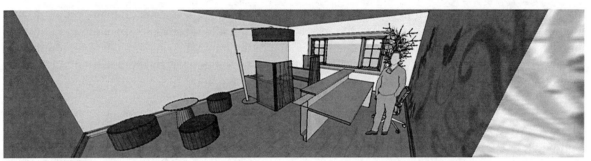

FIGURE 18.10 Third view adjusted

Now you will change the style applied to the new view.

21. Click on the **Styles** tab, within the *SketchUp Model* area (*Tool Tray*).
22. Click the small "**Home**" icon near the top.

23. Select **Styles** from the list.

24. Open the **Style Builder Competition Winners** folder.

25. Select **Pencil on light brown.**

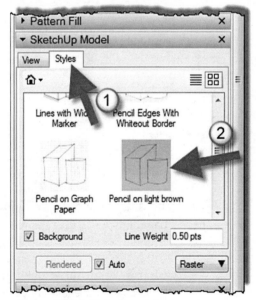

FIGURE 18.11 Changing style

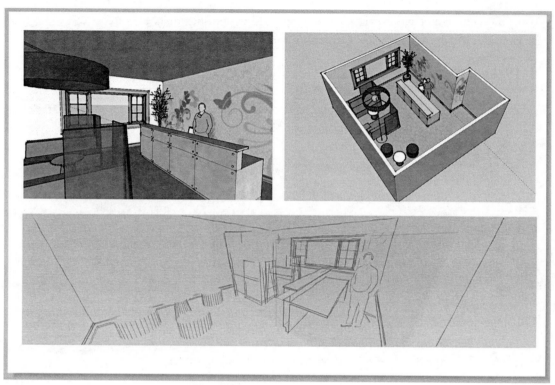

FIGURE 18.12 Different style applied to third view

Unfortunately, the style has sections turned on so the section we added to the original desk model appears. We would have to edit the style in the SketchUp model to fix this. We will ignore it. The last thing you will do to this sheet is add text.

26. Select the text tool icon.

27. Click and drag to define a box in which to type your text. Do this for each image, enter the text shown in Figure 18.13.

With text selected, you can use the arrows keys on your keyboard to quickly move the text object around (this also works with selected views). If you need to do back and change text, double-click on the text to gain access to the text box.

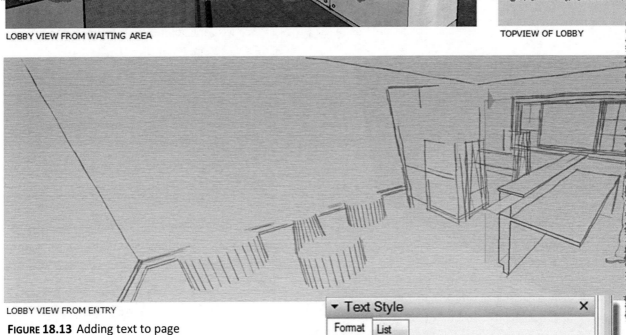

LOBBY VIEW FROM WAITING AREA

TOPVIEW OF LOBBY

LOBBY VIEW FROM ENTRY

FIGURE 18.13 Adding text to page

With the text selected (i.e. highlighted) you can adjust this properties in the **Text Style** window via the *Tool Tray* (Figure 18.14).

You have now completed the layout of a presentation sheet. This sheet can now be printed to a printer or to a PDF (if you have a PDF printer driver).

The next sheet you create in this document will be a construction-type document sheet.

28. **Save** your file as **Lobby Presentation**.

29. **Close Layout.**

FIGURE 18.14 Text Style panel

Now that we have looked at using *LayOut* for presentation drawings, we will now turn our attention towards the more format technical document approach. First we will open the model in *SketchUp* and setup a few Scene tabs to streamline the process once we get back into *LayOut*.

30. **Open** your lobby model in SketchUp.

31. Select the **Scene 1** tab to make it current (if not already).

We are about to create an elevation view of the front of the reception desk. In this case we want to see the ceiling. We will unhide it now.

32. Select **Edit → Unhide → All**.

The ceiling is now visible. If you were to click the *Scene 1* tab the ceiling would disappear again as we have not changed the current settings for the scene—which is good. Next we will add a section plane to hide the portion of building in the foreground of our desired elevation view of the desk.

33. Select **View → Section Cuts**; only if it does not have a check next to it.

34. Click on the exterior face of the left-hand wall as pointed out in Figure 18.15.

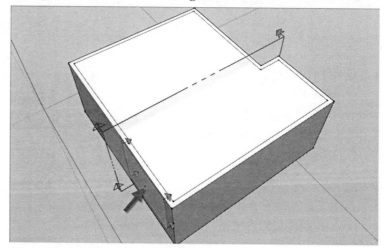

FIGURE 18.15 Adding new section plane

35. **Select** the new section plane and use the **Move** tool to reposition it as shown in Figure 18.16; just in front of the desk.

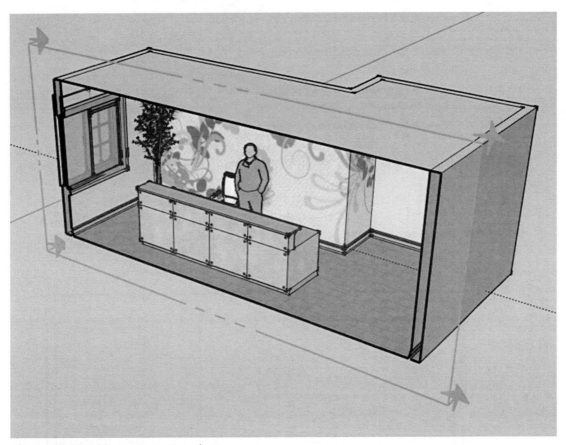

FIGURE 18.16 Adding new section plane

Seeing as we do not need any other sections we can clean house now.

36. Select the section in the other direction, if you have one, and delete it.

37. Right-click on the desk, **Edit Component**, and delete the section plane and then right-click **Close Component** to finish.

Next we will reposition the view to be an elevation.

38. Select the **Front** icon on the *Views* toolbar.

We also need to switch from perspective to parallel projection so the view looks like a 2D drawing.

39. Select **Camera → Parallel Projection**.

40. From the **Styles** window, select **Default Styles → Engineering style**.

41. Select **View → Section Plane**.

42. Select **View → Face Style → Wireframe**.

43. **Right-click** on the plant and select **Hide**.

Your view should now look similar to the image below (Figure 18.17).

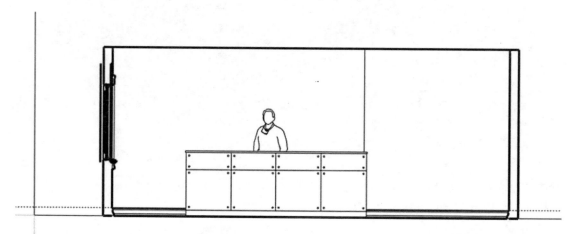

FIGURE 18.17 Elevation view of reception desk

44. Select **View → Animation → Create Scene**.

45. Select **Create Scene** from the prompt.

46. **Right-click** on the new tab and select **Scene Manager** from the menu.

Here you can rename the scene tab to make it more intuitive, both in *SketchUp* and *LayOut*. Also, notice the various properties saved with each scene and the fact that they do not have to be saved with the scene.

47. In the *Scenes* window, chance the *Name* to **Reception Desk**; click in the description field to apply the new name.

We will also create a floor plan view with similar settings. We can quickly generate this new view, starting from this view.

48. Select the **Top** view icon from the *Views* toolbar.

49. Select **View → Section Cuts**; to toggle them off.

50. **Right-click** somewhere in the middle to highlight/select the ceiling face.

51. Select **Hide** from the menu.

52. Create a new **Scene tab** and name it **Floor Plan**.

You should now have a floor plan view (Figure 18.19) created and ready for *LayOut*.

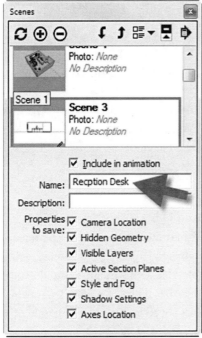

FIGURE 18.18 Scenes window

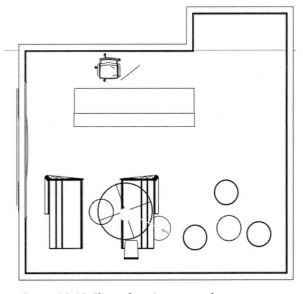

FIGURE 18.19 Floor plan view created

Now we are ready to get back into LayOut and set these new views up on sheets.

53. Save your *SketchUp* model.

54. Select **Send to LayOut**.

LayOut is now opened and you are prompted to select a template. Notice SketchUp does not know anything about the other *LayOut* document you created. However, *LayOut* does keep track of the *SketchUp* model as previously discussed.

55. For the template, select **Titleblock → Contemporary → ArchD Landscape**.

Arch D is a 24"x36" sheet of paper. This larger paper size allows larger scale drawings and/or more drawings per sheet. You need access to a plotter or print service to be able to print this size drawing. To use a print service, you would typically print the drawings to a PDF file and email or upload it. Unfortunately, the default templates do not offer 22"x34" which is generally preferred. This size allows for half-size drawing sets to be printed on 11"x17" paper. Most design firms have high-speed laser printers capable of print 11"z17" sheets much faster than a plotter.

56. Click **Open** to select the template.

You now have a new *LayOut* document with your current SketchUp view, the floor plan, positioned within a titleblock (Figure 18.21). The first thing you need to do is set the scale for the floor plan view as it was just scaled to fit the available area.

57. Right-click on the floor plan drawing, and select **Scale → 1/2" = 1'-0"** (Figure 18.20).

Notice how the "current scale" was listed in tis relative location. Often, you want the drawing to be as large as possible. This helps to take some of the guesswork out of which scale to select. We could have gone with the ¾" scale, but ½" is a plenty large floor plan.

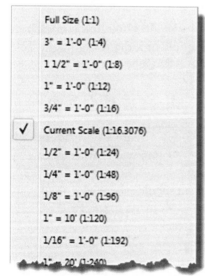

FIGURE 18.20 Scale list in LayOut

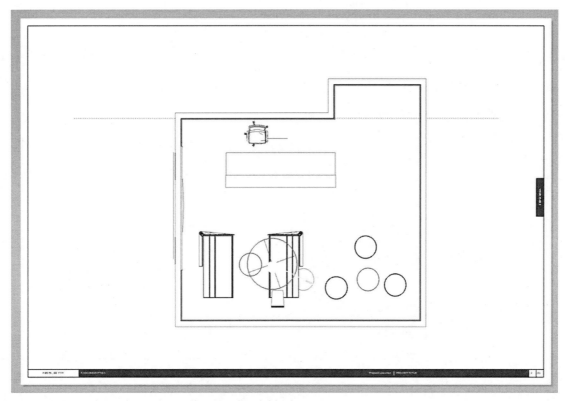

FIGURE 18.21 Floor plan view on sheet with titleblock in LayOut

With the scale set you can now readjust the viewport, which defines the perimeter of the view. This could even be used to crop the drawing if you only wanted to focus in on the desk area.

58. **Drag the viewport** so it is closer to the extents of the drawing.

59. **Move** the drawing to the upper right of the sheet as shown in Figure 18.22.

60. **Copy** the viewport to the right and change the scene to **Scene 1**.

61. Add the text shown below each drawing.
   a. *Title:* Verdana, Bold, 36pt
   b. *Scale:* Verdana, Regular, 24pt

62. Use the **Line** tool to draw a line below the title as shown

*Tip: Hold **Shift** for an orthogonal line.*

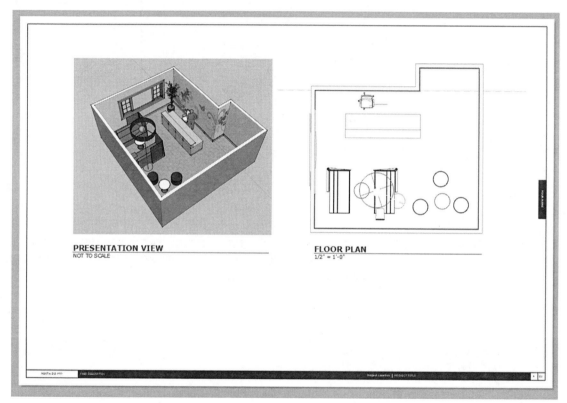

FIGURE 18.22 Presentation view and text added to sheet

Next you will create another sheet in this drawing set. This sheet will have the reception desk elevation on it.

63. Right-click on the **Inside Page** under *Pages* (Figure 18.23).

64. Select **Duplicate**.

65. Switch the floor plan view to be the elevation view; via the *SketchUp Model* window panel in the *Tool Tray*.

66. Move the elevation, text and line to the upper right of the sheet.

67. Select the presentation view on the left and delete it. Do NOT delete the text and line.

FIGURE 18.23 Creating new page

On this page we might want to have the interior perspective view of the lobby. If you rendered it in the previous chapter, using V-Ray for SketchUp, you can insert a raster image.

68. From the **File** menu, select **Insert**.

69. **Browse** to the raster image (i.e. JPG or PNG) of you lobby rendering.

   TIP: *If you don't have a rendering file, just select any raster image so you can practice this process.*

70. Place the image and resize approximately as shown in Figure 18.24.

Working with Text, Dimensions and Patterns

The last thing we will look at is adding text, dimensions and patterns to these views. As you have already seen, it is possible to add text and dimensions in *SketchUp*, but it is a bit tricky. *LayOut* makes this process much easier as you are working with fixed views and don't have to worry about things showing up in other views.

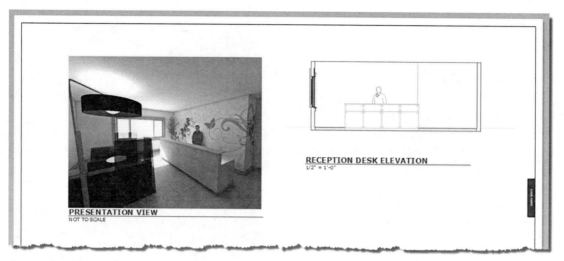

RECEPTION DESK ELEVATION
1/2" = 1'-0"

PRESENTATION VIEW
NOT TO SCALE

FIGURE 18.24 Second sheet with views positioned

Let's start by adding dimensions to the floor plan.

71. Switch to the first page; select **Inside Page** from the *Pages* panel.

72. Zoom in to the floor plan drawing.

73. Select the dimension icon from the toolbar.

74. Add the dimensions shown in Figure 18.25.

> ***FYI:*** *Pick three points; the first two are what you want to dimension and the third is the location of the dimension line.*

> ***TIP:*** *Be careful to pick the wall line and not the lines within the floor base.*

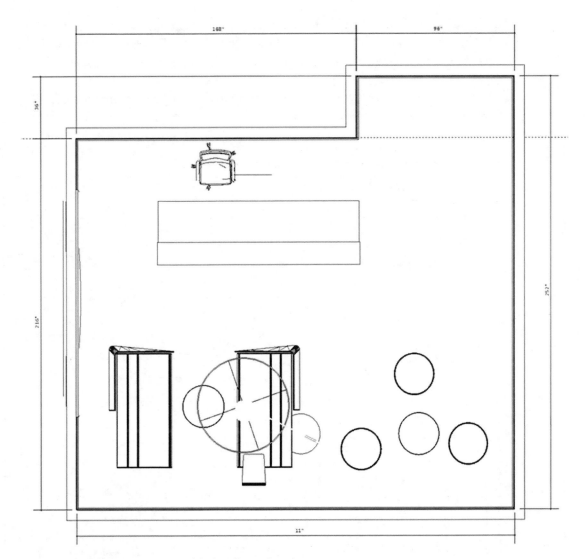

FIGURE 18.25 Dimensions added to floor plan view

75. Add the text and labels show in Figure 18.25 below. You have already used the text tool. The label tool is similar to text but includes a leader. Your first pick, for the leader tool, is near what you want to point at.

TIP: Placing a label involved picking two points and typing text. If you click drag one OR both of those points you can create curved leaders.

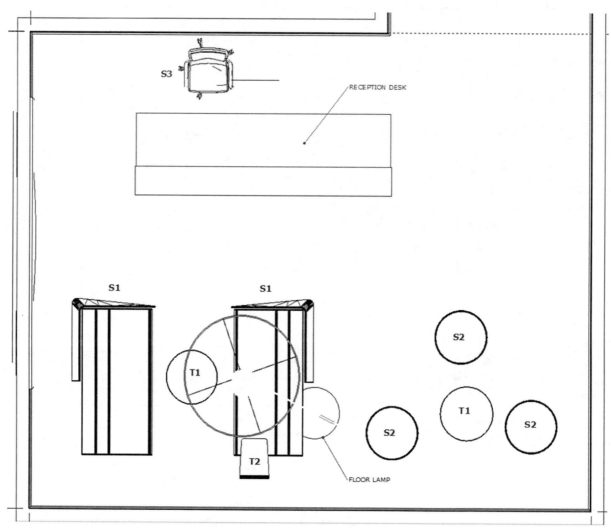

FIGURE 18.26 Text and labels added to floor plan view

Next you will fill the walls with a pattern. This can help to denote materials and make for more graphically readable drawings.

76. In your floor plan view, zoom in on the upper-right corner of the space.

To add a pattern you need to create an enclosed area using any combination of the line, arc, rectangle, circle, polygon, split and join tools. We will work on one section of wall at a time.

77. Add the *Pattern* per the steps below (Figure 18.27).

- Using **Rectangle** tool, select two points (Steps 1-3)
- Click the **Selection** tool and the select the rectangle (4 & 5)
- Click the **Pattern** button, click the pattern swatch to change pattern (6 & 7)
- Select the **Steel** pattern from the *Material Symbols* pattern group (9)

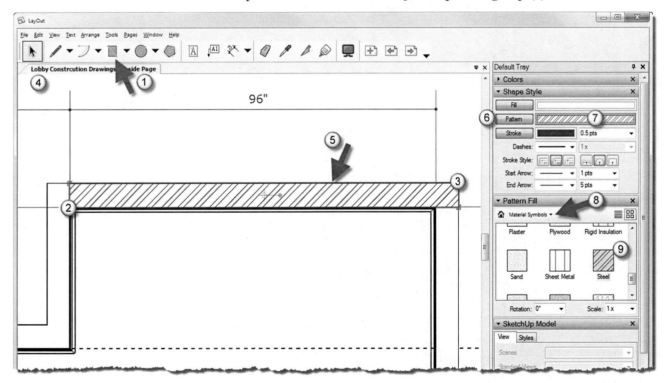

FIGURE 18.27 Adding a pattern to the walls

78. Repeat this process for all the walls. Do not cover the windows. Suggest a door opening in the east wall as shown (Figure 18.28).

79. **Save** your project as **Lobby Construction Drawings**.

Adding patterns in *LayOut* is a relatively new feature. Hopefully this feature will allow us to simply pick in an enclosed area and find the boundary in the future. It would be nice if the boundary could easily be made invisible as well. This concludes our introduction of *LayOut*.

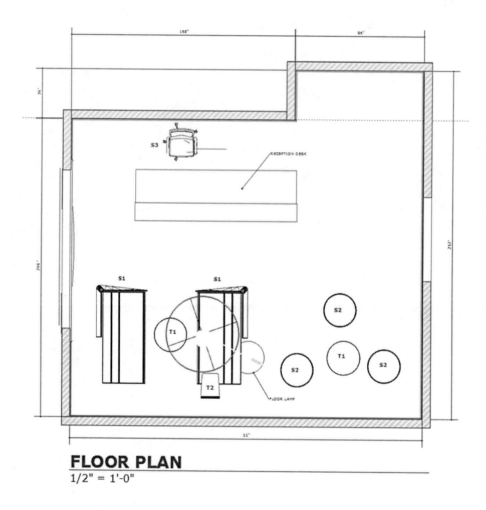

FLOOR PLAN
1/2" = 1'-0"

FIGURE 18.28 Pattern added to walls

Chapter 18 Review Questions

The following questions may be assigned by your instructor as a way to assess your knowledge of this chapter. Your instructor has the answers to the review questions.

1.   LayOut only comes with SketchUp Professional. (T/F)

2.   LayOut can be used to created construction documents. (T/F)

3.   Text and Dimensions may be added to the drawing in LayOut. (T/F)

4.   Each viewport must be the same drawing scale on a given sheet. (T/F)

5.   Interior elevation views are set to *Parallel Projection*. (T/F)

6.   Fill patterns may be added to embellish your drawings as needed. (T/F)

7.   The saved *Scenes* from the SketchUp model are accessible within LayOut. (T/F)

8.   You cannot have a titleblock on a sheet. (T/F)

Chapter 19
Working with AutoCAD DWG Files

It is fairly common to start a SketchUp model using an existing AutoCAD, DWG format, file. For years now, AutoCAD has had the largest market share for architectural design software. Therefore, there are a lot of existing buildings already drawn in AutoCAD. It is WAY faster to start with an AutoCAD file than it would be to sketch everything from scratch. It can take several days to accurately draw the existing conditions of a building, so having access to a DWG file is a big plus.

This chapter will walk you through the process of using a DWG file as the starting point for a SketchUp model. Most DWG files are 2D drawings; typically, one file for each floor of the building. SketchUp can import either a 2D or 3D AutoCAD file.

One more thing to point out before we get started... this feature, the ability to import a DWG file, is only available in the **Pro** version of SketchUp. **The free version of SketchUp cannot import AutoCAD DWG files.** This is probably the main reason most people buy the professional version. As was just mentioned, this feature often saves days' worth of work, so the cost is worth it. One nice thing about the *Pro* version for design firms is they can buy network licenses. If a company buys five network licenses, they can install SketchUp on one hundred computers, but only five people can be using the software at any given time. They can still buy standalone licenses, too; this is one copy for each individual computer.

> *FYI: This chapter requires* **SketchUp Pro**, *not the free version used in the rest of this book. Also, you will have to download the resource files from the publisher's website* (www.SDCpublications.com) *to follow along exactly. Otherwise, you can use another AutoCAD DWG floor plan file if you have one.*

1. Open a new SketchUp file.

2. Select **File → Import**.

> *TIP: If you do not have access to SketchUp Pro, you can still complete this section by downloading the file* **Library.SKP** *from* www.SDCpublications.com. *This file already has the DWG file loaded into it. Once the DWG file is in the SketchUp model, you no longer need the Pro version.*

3. Browse to the AutoCAD DWG file **Library – Level 01.dwg**; <u>do not</u> click open yet.

4. Click the **Options** button (Figure 19.1).

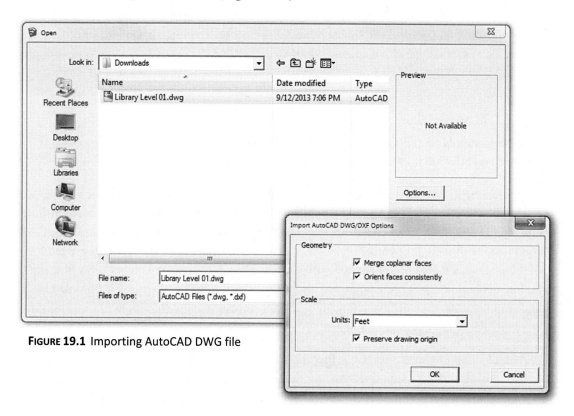

FIGURE 19.1 Importing AutoCAD DWG file

In most cases you will want both options checked, under geometry, to make sure you start with a clean and organized model. You will also check "Preserve drawing origin". This will align the AutoCAD drawing origin (0,0,0) with the SketchUp origin, which is the intersection of the *Axes* lines. This is nice in case you continue to receive updated/progress drawings, as they will always land in the same location. Otherwise you have to pick a place to insert the DWG.

5. In the *Options* dialog, make sure **all the boxes are checked** and *Units* are set to **Feet**.

6. Click **OK**.

7. Click **Open** to place the DWG file in SketchUp.

8. Take a minute to note the entities listed and then click **Close**.

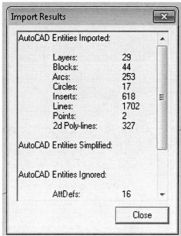

FIGURE 19.2 Import results

You should now see the first floor library plan as shown in the image below (Figure 19.2). Notice how the SketchUp person, near the origin, looks to be about the right size in relation to the imported drawing. This is a good clue that the model is the correct scale.

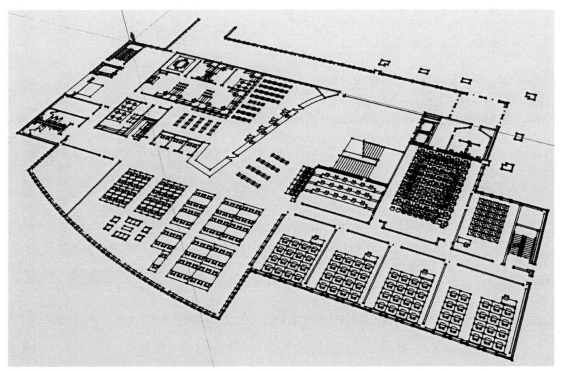

FIGURE 19.3 Imported AutoCAD DWG file

Next you will use the *Tape Measure* tool to verify the drawing came in at the correct scale.

9. **Zoom in** on a door opening into one of the classrooms in the Southeast corner of the building.

10. Use the **Tape Measure** tool to verify the doors are **3′-0″** wide (Figure 19.4).

   FYI: The scale should be correct for this example, but when it is not, you will need to use the Scale *tool to make it right.*

Next you will open the *Layers* window and notice several have been created. The AutoCAD user typically uses layers to control visibility and other properties of entities.

11. Select **Window → Layers**.

You should see the layers as listed in Figure 19.5. We will turn off *Layer* "A-Flor-Nplt" to hide the clearance lines built into the furniture. One might also want to turn off other *Layers* to reduce the visible lines to just walls, casework and windows.

12. Uncheck **A-Flor-Nplt**.

In the Layers window, the *Layer* with the circle selected, in the left-hand column, is considered the current *Layer*. This is the *Layer* on which all new edges and faces are created on.

13. **Close** the *Layer* window by clicking the "X" at the upper right.

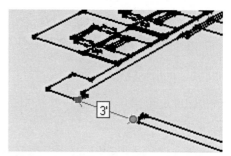

FIGURE 19.4 Tape Measure tool

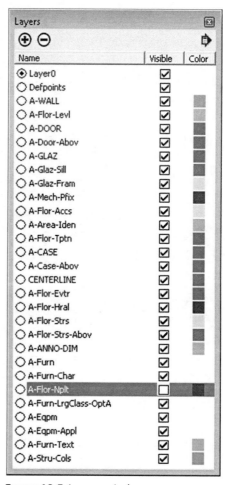

FIGURE 19.5 Layers window

Next, you need to decide if you want to **Explode** the DWG file or not. Right now, if you select the imported drawing, the entire thing highlights. Opening *Entity Info* shows that this is a *Component* (Figure 19.6). This is nice if we want to move it around, we don't have to worry about selecting hundreds of files and text. Or, if we want to replace it with an updated DWG file.

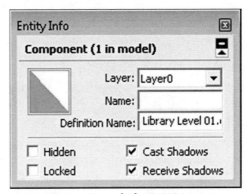

FIGURE 19.6 Entity Info for DWG Import

However, if we want to edit the lines or extrude them into the third dimension we will need to *Explode* the *Component*. In the next step you will explode the floor plan drawing.

14. Right-click on the floor plan drawing.

15. From the pop-up menu, select **Explode**.

Now, when you select a line, you will just be selecting that one line, not the entire floor plan.

After exploding the drawing, you will just have lines and no faces. To create faces you can sketch a line diagonally at a corner or on top of another line. Another option, when using the Pro version is to use a **Ruby Script**. This feature allows things to be automated in SketchUp. Here is one example of a tool which will create faces for you: **www.smustard.com/script/MakeFaces**.

Once you have faces you can follow the steps outlined in the previous chapter to create a 3D model!

16. **Save** your model as **Library.SKP**.

FIGURE 19.7 Right-click menu

Chapter 19 Review Questions

The following questions may be assigned by your instructor as a way to assess your knowledge of this chapter. Your instructor has the answers to the review questions.

1. SketchUp and AutoCAD are made by the same company. (T/F)

2. *Layers* can be used to control element visibility. (T/F)

3. *Entity Info* dialog will show us which *Layer* a selected element is on. (T/F)

4. SketchUp Make, the free version, can import DWG files. (T/F)

5. SketchUp recognizes the layers in the DWG file. (T/F)

6. The DWG import cannot be *Exploded* down to editable SketchUp elements. (T/F)

7. During import, you can specify the units for the DWG file (e.g. feet or inches). (T/F)

8. Once a DWG is imported, use the *Tape Measure* tool to verify the scale. (T/F)

Chapter 20
Working with Autodesk® Revit® Files

This chapter will discuss the ways in which *Autodesk Revit* models can be used within *SketchUp* and vice versa. This author has worked on a number of projects which required some level of file sharing between these two applications. We will start with a few examples to better understand why you would want to do this in the first place. Once we understand why, we will then work through two examples.

SketchUp → Revit

- A preliminary design was developed in SketchUp through the Schematic Design (SD) phase. As the project moves into Design Development (DD) the SketchUp model is imported into Revit and used as a base when creating the Building Information Modeling (BIM) elements.

- A university student center was being developed in Revit. The design firm had developed a complex multi-level stair and glass railing model in SketchUp. Given the some of the limitations within the stair and railing tools in Revit, and their staff knowledge, it was decided to use the SketchUp stair model in Revit.

Revit → SketchUp

- The interior designer of a lobby, for a healthcare project, wanted to bring the Revit model in SketchUp and quickly create a sketchy, more artistic, presentation and animation of the space. Revit cannot create the sketchy line work like SketchUp, and creating animations are a little more involved.

- An addition is proposed on a building previously modeled in Revit. The 3D Revit model is loaded into SketchUp so the designer can quickly explore different options with a less refined sketchy look.

These scenarios may not be the ideal workflow in every situation, but having an assortment or tools in your "toolbox" is beneficial to any designer.

The following examples will require the use of *Autodesk Revit*. This tutorial is based on Autodesk Revit 2014. Similar to SketchUp, there is a trial version which may be obtained from www.autodesk.com.

Firs we will bring a Revit model into SketchUp. Revit cannot directly import the Revit file format (*.RVT). So you will need to open the file in Revit and export the model to a DWG file format. This is an Auto CAD file format, which SketchUp Professional can import. These next few steps are not meat to teach you specifically how to use Revit; you just need to follow the steps to export the model. If you do not have access to Revit, you can use the provided DWG; which is the end result of exporting the model form Revit.

1. Open the provided file **Revit Model.RVT** by double-clicking on it in *Windows Explorer*.

You should now see a 3D view of your Revit model (Figure 20.1). You must be in a 3D view before using the Export to CAD tool; otherwise you will get a 2D file. Notice the 3D view name is highlighted in the Project Browser.

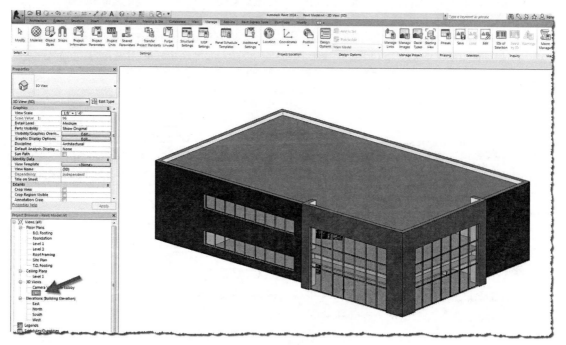

FIGURE 20.1 3D view of provided Revit model in Autodesk Revit 2014

2. With the 3D view open, select the **Application Menu**; the Purple "R" in the upper left corner.

3. Select **Export → CAD Formats → DWG** (Figure 20.1).

4. Click the "**…**" button to edit the export settings (#1; Figure 20.3).

5. Select **ACIS solids** on the *Solids* tab (Figure 20.4).

6. Click **OK**.

7. Click **Next**.

8. Change the *Files of type* to **AutoCAD 2010 DWG Files**.

9. Click **OK**.

10. Close Revit by clicking the red "**X**" in the upper right—do not save.

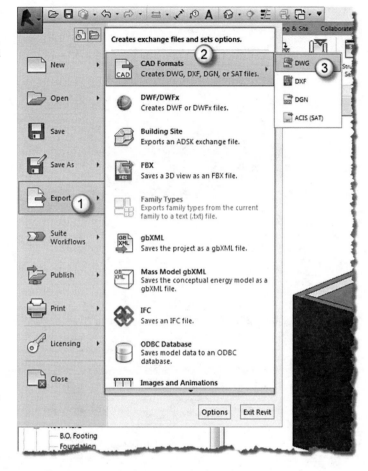

FIGURE 20.2 Autodesk Revit application menu

In some cases you might want to hide any extra geometry in Revit before exporting the file. Larger Revit projects can have some many 3D faces that SketchUp will run too slow. In a Revit view, simply type **VV** to see a list of categories that can be turned off. You can also select things, right-click and pick **Hide in View → Elements** from the pop-up menu.

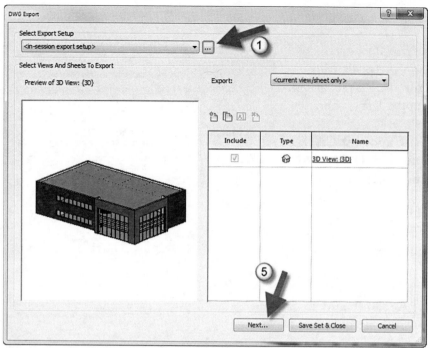

FIGURE 20.3 Autodesk Revit export to DWG dialog

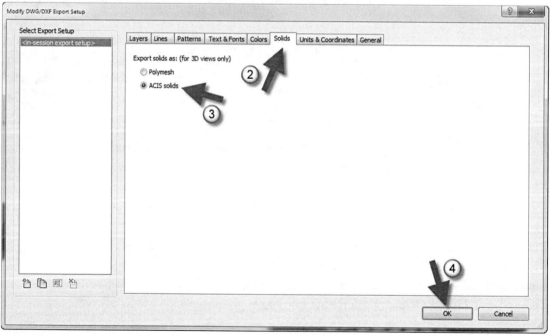

FIGURE 20.4 Autodesk Revit export to DWG options dialog

You are now ready to import the file you just exported into SketchUp.

11. Start a new, empty project in SketchUp.

12. Select **File → Import**.

13. At the bottom of the *Open* dialog, change the **Files of Type** to **AutoCAD Files (*.dwg, *.dxf)**. This requires the Pro version of SketchUp.

FYI: *This changes the filter to only look for files with this specific extension.*

14. **Browse** to the location you saved the DWG file, and then **select** it.

15. Click the **Options** button.

16. Click **OK** without making changes in *Options.*

17. Click **OK** to import the DWG.

18. Click **OK** to the Import Results (Figure 20.5).

19. Click the **Zoom Extents** icon to see the entire model.

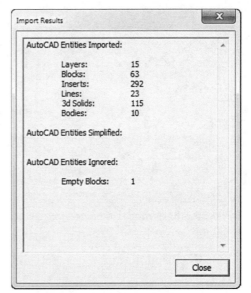

FIGURE 20.5 DWG import results

You should now see the building model in SketchUp (Figure 20.6). At the moment SketchUp is treating this new import as a Component; the same as it did when you imported the reception desk into the lobby drawing. You could "edit component" every time you needed to make changes to this model, but it might be easier to **Explode** in this case.

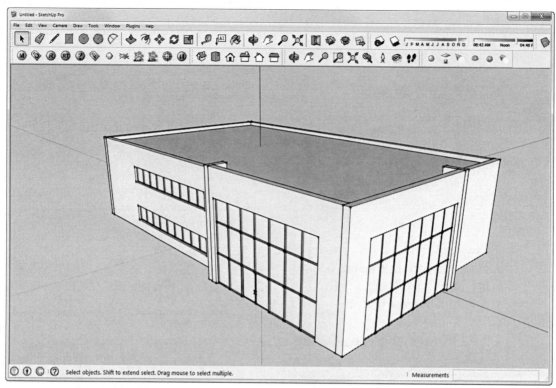

FIGURE 20.6 Revit model imported into SketchUp as DWG file

20. **Right-click** on the drawing and select **Explode** from the context menu.

Because this DWG file had multiple *Blocks* (in AutoCAD this is similar to a *Component*) it will have to be *Exploded* multiple times. You could leave some things as *Components*, but for simplicity we will explode everything.

21. Select **Edit → Select All**, and then right-click on the model and select **Explode**; Repeat this step until Explode is no longer an option (i.e. it is grayed out).

22. Use the **Paint** tool to add **Brick, Corrugated Metal and Glass** to the exterior as shown in Figure 20.7.

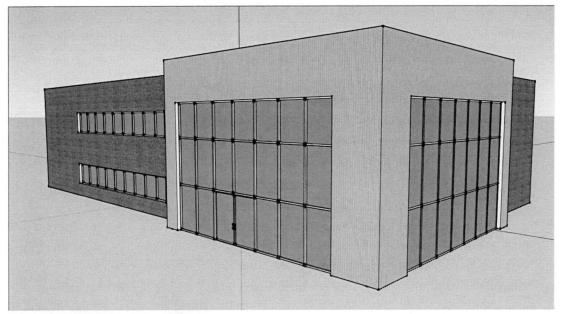

FIGURE 20.7 Adding materials to exterior

Because the Revit model has the exterior glass modeled as 1" thick. The SketchUp model now has two faces in each window opening. So adding the glass material to the one face does not allow you to see through the building to the interior. We will deal with that next.

23. Right-click on one of the glass faces.

24. Pick **Select → All on same layer.**

The DWG export from Revit is very detailed in how it is separated onto SketchUp Layers. This will allow us to quickly paint all layers,

25. Now **Paint** the glass material again.

You should now be able to see through all the glass (Figure 20.8). Sometimes you might need to "select all on layer" and then right-click and select **reverse faces** so you can paint both sizes of the face.

At this point you can start creating Scene tabs, and working your way into the model. Adding materials and loading SketchUp based content. Just to practice this, you will load your reception desk into the lobby area of this building.

26. Select **Tools → Section Plane**.

27. Click somewhere on the roof to create a horizontal section plane element.

28. Use the **Move** tool to reposition the *Section Plane* as shown in Figure 20.9.

29. As previously covered, use the **File → Import** tool to bring in your reception desk. Place as shown in Figure 20.9.

    TIP: *Be sure the "Files of type" is set to SKP, as it may still be set to DWG.*

30. Edit the reception desk *Component* and delete the section plane if it is still there.

31. Create a **Scene** tab named First Floor.

32. Turn off the **Section Cut** and **Plane** via the View menu.

33. **Save** the *SketchUp* file as Small Office Building.skp.

FIGURE 20.8

Glass material added to all faces of glazing

You need to save the file as an older SketchUp version for Revit to load it properly.

34. Select **File → Save As**, change the *Save as type* to **SketchUp version 8**; name the file Small Office Building V8.skp.

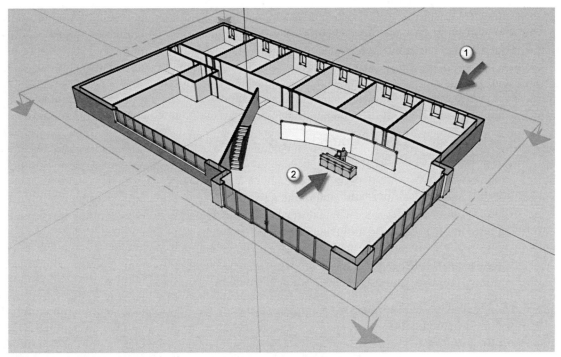

FIGURE 20.9 Reception desk added to first floor lobby

Now you will repeat this process in the opposite order. You will take the SketchUp model you just saved and import it into Revit. Revit can read the SketchUp format directly so there is no need to export it to another format first. However, you had to save it as an older version of SketchUp.

35. Open Autodesk Revit 2014.

36. Select **Application Menu → New → Project** (Figure 20.10).

37. Pick the **Architectural Template** from the drop-down list.

38. Click **OK**.

You are now in a new/empty Revit project, currently in the Level 1 view (see the *Project Browser* list). You can import the SketchUp model into this view.

Before you start the Import command, there is one thing you need to do first. In order to get the SketchUp model to cut properly in plan and section views we need to import the model into what is called an **In-Place Model**. This is similar to the idea of a component in SketchUp.

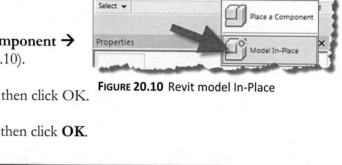

39. Select **Architecture → Component → Model In-Place** (figure 20.10).

FIGURE 20.10 Revit model In-Place

40. Select Generic Models and then click OK.

41. Name it <u>SKP Building</u> and then click **OK**.

You are now in a temporary edit mode for this *In-Place Model*. **The only way to end it is by clicking the red X or the green check-mark on the *Ribbon*.**

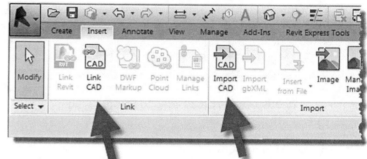

You now have one of two options on the Insert tab; *Link* or *Import* (Figure 20.11). The best option is usually *Link*. If any changes are made to the SketchUp model, the

FIGURE 20.11 Revit insert tab

42. Click **Link CAD**.

43. Change the *Files of type* to **SketchUp (*.skp)**.

44. **Browse** to your reception desk file and **select it**.

45. Click **Open** to load it.

46. Click the **green check mark** to finish the *In-Place Model* (#1; Figure 20.12).

47. Click the **Default 3D View** icon on the *Quick Access Toolbar* (#2; Figure 20.12).

48. Change the *Visual Style* to **Consistent Colors** (#3 & 4; Figure 20.12).

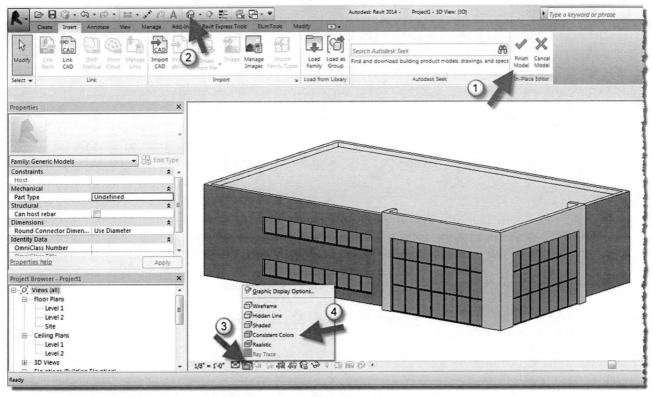

FIGURE 20.12 SketchUp model linked into Autodesk Revit

Before we finish, let's look at a couple of things you should know about controlling how the SketchUp model looks in Revit.

49. Type **VV** on the keyboard (do not press Enter).

50. Click on the **Imported Categories** tab (Figure 20.13).

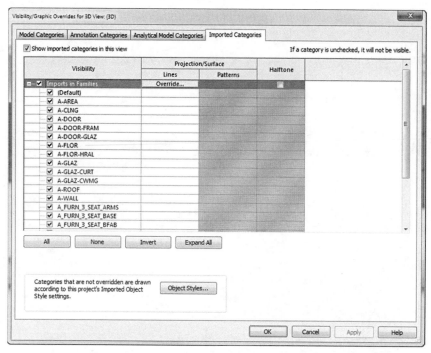

Here you can control the visibility of some elements within the SketchUp model. You can see similar settings for project-wise options via **Manage** (tab) → **Objects Styles** → **Imported Objects**.

51. Click **OK** to close the dialog.

Next you will make the glass transparent.

FIGURE 20.13 View settings for Revit 3D view

52. Select **Manage** (tab) → **Materials** (button) from the *Ribbon*.

53. Select the Revit material named **Render Material 100-149-237** (Figure 20.14).

54. Change the *Transparency* setting to **50** (#2; Figure 20.14).

55. Click **OK**.

The glass is now transparent (Figure 20.15). You have to select each material to see what color it is, and then compare that to what you see in the model to determine which Revit material to edit. Zoom in to see the reception desk and a message from the "SketchUp Guy"!

56. **Save** your Revit model as Linked SKP Model.rvt.

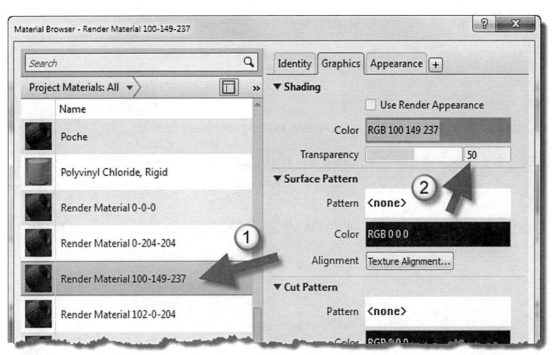

Figure 20.14 Revit materials created from SketchUp model link

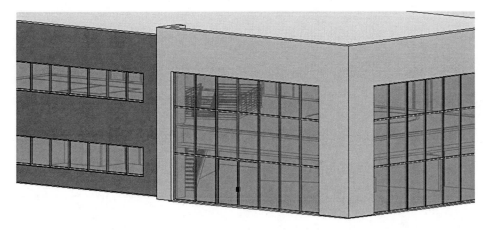

Figure 20.15 Glazing from linked SketchUp model made transparent in Revit model

Sample project courtesy of LHB, Inc. and Hybird Medical Animation

Chapter 20 Review Questions

The following questions may be assigned by your instructor as a way to assess your knowledge of this chapter. Your instructor has the answers to the review questions.

1.   SketchUp can import a Revit RVT file. (T/F)

2.   Revit can import a SketchUp SKP file. (T/F)

3.   Autodesk provides a free 30-day trial for Revit. (T/F)

4.   The imported DWG file needs to be Exploded before applying materials. (T/F)

5.   When importing an AutoCAD file, be sure the *Files of Type* is set to DWG. (T/F)

6.   The glass was not initially transparent because there are multiple layers (i.e. faces) representing glass. (T/F)

7.   Revit can export a DWG, which SketchUp can then import. (T/F)

8.   In Revit, the SketchUp model needs to be importuned into an **In-Place** family to look correct in plan and section views. (T/F)

Additional Resources

This is by no means an exhaustive list, but here are a few resources you may want to check out if you are interested in learning more and keeping up on the latest SketchUp info:

- **Within SketchUp**
  - **Help → Welcome to SketchUp** and then click on the "**Learn**" bar for links to *Video Tutorials* and *Tips & Tricks*
  - Don't forget about the **Knowledge Center** (aka Help Center)
  - **Window → Instructor**

- **Online**
  - YouTube
    http://www.youtube.com/profile?user=SketchUpVideo&view=playlists
  - SketchUp.com
  - http://www.sketchupartists.org/

- **Blogs**
  - http://sketchupdate.blogspot.com/
  - http://sketchuptips.wordpress.com/

- **Components**
  - 3D Warehouse, via…
    - File → 3D Warehouse → Get Models
    - http://sketchup.google.com/3dwarehouse/
  - Plumbing Fixture Manufacturer example:
    http://sketchup.google.com/3dwarehouse/
    cldetails?mid=886afdb19cf5a0e12e8ae34ac2aaf4f2
    Furniture manufacturer example:
    http://www.hermanmiller.com/design-resources/3d-models-revit/3d-models-by-product/seating.html

Index

The images below are from a sample project provided with this book. The project was designed by LHB Inc. (www.LHBcorp.com) and modeled by *Nick Vreeland.*

The images on the left are native SketchUp views. On the right is the same view rendered in V-Ray for SketchUp; an add-in app covered in chapter 17.

Project: Hybrid Medical Animation, Minneapolis MN

Native SketchUp view – Open Office

Rendered V-Ray for SketchUp view – Open Office

Native SketchUp view - Conference Room

Rendered V-Ray for SketchUp view - Conference Room